UAS Integration into Civil Airspace

Aerospace Series

UAS Integration into Civil Airspace

Policy, Regulations and Strategy

Douglas M. Marshall

Registered Offices
John Wiley & Sons, Inc., 111 River Street, Hoboken, NJ 07030, USA
John Wiley & Sons Ltd, The Atrium, Southern Gate, Chichester, West Sussex, PO19 8SQ, UK

Editorial Office
111 River Street, Hoboken, NJ 07030, USA

For details of our global editorial offices, customer services, and more information about Wiley products visit us at www.wiley.com.

Wiley also publishes its books in a variety of electronic formats and by print-on-demand. Some content that appears in standard print versions of this book may not be available in other formats.

A catalogue record for this book is available from the Library of Congress

Hardback ISBN: 9781118339497; ePub ISBN: 9781118536582; ePDF ISBN: 9781118536599; OBook ISBN: 9781118536575

Cover image: © windjunkie/Getty images
Cover design by Wiley

Set in 9.5/12.5pt STIXTwoText by Integra Software Servives Pvt. Ltd, Pondicherry, India

SKYEA9DFA29-AF48-4A29-8852-BEF6FD300B2F_031622

Contents

powered aircraft are in the airworthiness approval stage with the FAA. Urban air mobility concepts under development by companies such as Airbus, Joby Aviation, Kitty Hawk, Lilium, Terrafugia, and Uber Air are well on their way to securing civil aviation authority approvals to market eVTOL (electric vertical takeoff and landing) prototypes that can operate at low altitudes in urban environments. "Self-flying" air taxis are already taking passengers on test flights in Chinese cities. Researchers, developers, regulators, and others are working very hard to create remotely piloted systems that can share airspace at lower airspace levels, in an environment that is unlikely to confront a remotely piloted B777 loaded to the rafters with computer chips and smart phones, and flying at 400 feet above ground at 250 kts. What is now possible is a remotely or autonomously piloted aircraft carrying emergency medical supplies to a person in need, dispatched from the nearest public safety facility and operating in your neighborhood at an altitude of less than 250 feet above ground level (AGL), possibly saving a life. Or delivering that FedEx package that may have been carried to a distribution center by a conventional aircraft, but replacing the ground vehicle currently needed to complete the last leg of the journey. While technologically possible now, these scenarios may only take place outside of strictly controlled test environments under the watchful eyes of regulatory agencies such as the US Federal Aviation Administration or its functional equivalents elsewhere, national civil aviation authorities.

There are many moving parts in the realm of unmanned aircraft systems. Capturing all of them and doing them justice would require several volumes. The one essential component of the UAS "big picture" is airspace management; thus the focus of this book. Even that subcategory calls for a recursive analysis, as any technology is made up of many components that themselves are technologies, which have subparts that are also technologies, and so on, in a repeating, or recurring fashion. The airspace management function is typically the exclusive province of civil aviation authorities focused on safety and the fundamental goal of keeping aircraft separated from one another so as to not create a hazard of a mid-air collision. This function has been largely successful for over 60 years, depending upon the country in question. Generally, the denser the airspace traffic, the greater the likelihood of a mishap. Midair collisions, though relatively rare when compared to the number of aircraft in flight at any one time in congested airspace, such as parts of the US and Europe, still happen, often with tragic results. In the list of the top 10 leading causes of fatal general aviation accidents in the US from 2001 to 2017, the last year this statistic is available, midair collisions ranked number eight. The number of near-midair collisions reported each year is approximately 200, and actual collisions average between 15 and 20. General aviation hours flown (those most important to our analysis because they are more likely to be found at low operating levels), totaled 25.9 million in the calendar year 2019.

The theme of this book focuses on just one of those many moving parts, the integration of unmanned aircraft into controlled and uncontrolled airspace. The ongoing regulatory and policy efforts around the world to achieve full airspace integration will be examined, which requires a functional breakdown of the key elements of the technology that must meet regulatory requirements before the systems will be permitted to go into full operation. Predictions of the future in this technology sector are fraught with uncertainty, but an attempt will be made to outline a probable path forward as revealed by government regulators and myriad interested parties. Among the other moving parts are ongoing regulatory developments for Operations Over People, operations Beyond Visual Line of Sight of the pilot in command, Remote Identification of aircraft, and Night Operations. While these components are critical features of the overall challenge of airspace integration, they will not be discussed in detail in this volume, except to the extent that it is necessary to clarify their role in the integration picture.

The rapid evolution of the technology underlying unmanned aircraft, unmanned aircraft systems, remotely piloted aircraft systems, and, more commonly, drones, among other terms of art, presents a formidable challenge to anyone attempting to encapsulate the entire domain in one book. The broad notion of unmanned or remotely piloted aircraft has been with us for over a century. Society has witnessed extraordinary developments in the field of unmanned aviation over the last 30 years or so. The categories or topics that now define or bound the current state of the art of this technology are too numerous to list here, but will be discussed in the body of this book.

The media and popular press adopted the catchall label "drone," while experts and regulators generally prefer UAV, UAS, RPA, or RPAS instead of "drone," because the term "drone" once had a very specific meaning. The early descriptions of these types of aircraft settled on "drone," although the historical root of the term remains controversial (likened to a honeybee drone, or perhaps derived from a 1930s British target drone called a "Queen Bee"? No one really knows). In the earliest phases of development of unmanned aircraft, "drone" referred to target aircraft or remotely piloted offensive weapons deployed by both sides in both World Wars. Today's unmanned aircraft are far more sophisticated, and in most cases more capable of non-military missions than the "true" drones of nine decades ago. This book will use drone, UAS, sUAS, UAV, UAS, and RPAS more or less interchangeably, as a matter of style and continuity, unless the specific term used calls for an explanation of why it is used in the context of the discussion.

Military organizations, primarily in the United States, led the way in developing drone technology from flying targets for aircraft weaponry and surface-based artillery to aerial sensor systems modified for intelligence, surveillance, and reconnaissance (ISR) missions. That capability soon led to development of platforms capable of carrying and deploying highly effective offensive munitions (such as the General Atomics' Predator and Reaper series used in the first Gulf War and thereafter in subsequent Middle East conflicts).

Alongside the military's tactical and strategic adoption of unmanned aerial systems, and the emerging market for civilian or scientific versions of those systems, the consumer sector emerged, which quickly realized the potential for unmanned systems in both the recreational and commercial arenas. Small (weighing less than 55 lb, or 2.2k), fixed-wing and rotorcraft contrivances were soon adapted for aerial photography, agriculture, building and infrastructure inspection, package delivery, entertainment, and any number of other applications having nothing to do with military operations. The sudden "Cambrian Explosion" of affordable and highly capable (consumer based) small UAS soon overwhelmed civil aviation authorities around the world with demands for access to low level airspace for a multitude of civilian uses.

The growth of this technology has outpaced the ability of governing authorities at all levels, international, national, regional and local, to keep up with the changes and promulgate rules, regulations, and standards for the operation of these systems in the public domain, namely the airspace above the surface. As a consequence, rogue operators and abusers of the technology have created havoc with their misuse of consumer drones to invade privacy, disrupt wildlife, interfere with firefighting and law enforcement activities, endanger manned aircraft operations around airports, and any number of irresponsible uses of affordable and readily available off-the-shelf drones. The need for safe and predictable environments for legitimate users of this technology is paramount, and airspace integration strategies are likely to offer the most achievable solutions.

This book will only briefly address the history of this technology, as many other publications have covered the same ground, but will provide a framework for understanding and evaluating just one critical element of this extremely complex environment: how to integrate these systems, large and small, fast and slow, heavy and light, all without pilots on board, with other occupants of the airspace, namely manned aviation, and how to do it safely, equitably, and efficiently to minimize the risk of disaster and maximize the economic opportunities sought by the users of the airspace. The full solution to the safe integration challenge, which has eluded the experts, developers, and regulators thus far, is the key to the further advancement of the technology beyond its current status. The ongoing global efforts characterize the potential solutions to this challenge as Unmanned Aircraft Traffic Management (UTM), or Unmanned Aircraft Systems Space (U-Space), or Advanced/ Urban Air Mobility (AAM/UAM), depending upon in which part of the world the effort originates.

Organization of the Book

The following chapters will introduce the reader to the major issues confronting the developers of these strategies and provide a brief introduction to what each nation or group of nations is doing to address those issues. The templates adopted by the major contributors as they work their way through the often conflicting and sometimes overlapping regulatory environments in which they must operate to be successful are discussed in greater detail. There are many parallel efforts to identify a path to full integration of unmanned systems with manned aviation, and they do not all agree on the strategy or the architecture to make it so. For this reason, this book is unashamedly broad in scope in some respects and rather narrow in others. The goal is to identify a common way forward for the evolving UAS industry and the regulatory authorities that must enable and monitor its growth to ensure public safety and economic viability.

Chapter 1 "Background" introduces a thumbnail history of aviation regulations, derived in some respects from the ancient Law of the Sea. This chapter briefly summarizes the first attempts to regulate airplanes and their pilots in the UK, the creation of an international regulatory body (ICAO) in 1944 as a product of the Chicago Convention on International Civil Aviation, and then moves on to the present-day regulatory system, both national and international, that oversees all aspects of commercial and private aviation.

Chapter 2 "UAS Airspace Integration in the European Union" is a longer chapter, and attempts to cover historical and ongoing regulatory efforts in the European Union regarding unmanned aircraft operations and standards. The EU has been very busy adopting regulations for UAS that will apply across all of its Member States, and more recently embarked upon the concept of a "U-space" that is intended to integrate UAS/RPAS into the

European airspace by establishing a new concept for how the airspace can be managed while not disrupting existing commercial and general aviation activities.

Chapter 3 "ICAO" covers the International Civil Aviation Organization and its airspace integration activities with specific focus on remotely piloted aircraft systems.

Chapter 4 "UAS Airspace Integration in the United States" discusses airspace integration efforts in the United States, in coordination with Europe's EASA and other national aviation authorities.

Chapter 5 "Global Airspace Integration Activities" takes a look at UAS integration efforts in a few selected countries that are considered to be representative of similar efforts in a growing number of ICAO's 193 Member States.

Chapter 6 "The Role of Standards" examines the role of Standards Development Organizations (SDOs) in the development of regulations and best practices.

Chapter 7 "The Technology" discusses the various domains of the evolving UAS and UTM/U-space technology, and includes suggestions for a methodologies for conducting a risk assessment and functional decomposition of complex systems.

Chapter 8 "Cybersecurity and Cyber Resilience" offers a historical view of global cybersecurity failures and ties that history to current efforts to identify risks and defensive mechanism to ensure the security of aviation systems.

This is not an engineering text, nor is it a law book, but is a bit of a hybrid of both, focused on the study of one highly technical sector of innovation and economic growth from the proverbial "30,000 ft" view. The reader is cautioned, however, that the technology advances are extremely dynamic, and innovation, or "the next best thing," is almost a daily occurrence, so that accurately predicting the future is a fool's errand. As new challenges emerge, entrepreneurs and developers will step up to meet them, occasionally creating a new technology or new subset of existing technology that may not have existed even a year ago. The best we can do is to understand what is happening at this point in time and acquire the tools to respond to the breakneck pace of innovation in unmanned aircraft systems.

Aerospace Series Preface

The field of aerospace is multidisciplinary, covering a large variety of products, disciplines and domains, not merely in design and engineering but in many related supporting activities. The interaction of these diverse components enables the aerospace industry to develop innovative and technologically advanced vehicles and systems. The *Aerospace Series* aims to be a practical, topical, and relevant series of books aimed at people working in the aerospace industry, including engineering professionals and operators, engineers in academia, and allied professions such as commercial and legal executives. The range of topics is intended to be wide ranging, covering design and development, manufacture, operation and support of aircraft, as well as infrastructure operations and advances in research and technology.

Unmanned air vehicles are a growing and increasingly accepted part of the aerospace environment. Small UAVs equipped with appropriate sensors can carry out leisure, small industry and official roles in the visible and IR spectrum. As their use expands, unmanned air systems will inevitably become involved with, and potentially conflict with, manned vehicles – as has already been demonstrated by numerous encounters near airports. There will need to be new regulations to allow the co-existence of UAVs with GAS, rotary wing, regional and transnational operations. These new regulations could require changes to on-board navigation and proximity warning systems as well as to ATM practices and standards.

This book – *UAS Integration into Civil Aerospace* – explores the integration of unmanned aircraft into controlled and uncontrolled airspace. It provides a comprehensive overview of regulatory and policy efforts required to move towards full airspace integration, as well as the technology that must be developed and approved for full operation of UAV systems. It also addresses the critical questions of cybersecurity and cyber resilience as they relate to UAV airspace integration. The global ATM system depends heavily on electronic communications and interconnectivity, any interruption of which could lead to potentially catastrophic consequences.

With the rapid evolution of UAV technology, aviation regulators at international, national, and local levels have struggled to keep pace with appropriate rules and standards to ensure that UAV systems operate in shared airspace in a safe, equitable, and efficient manner. This book outlines a path forward that minimizes the safety risks while maximizing potential economic benefits for all users of the airspace. In line with the mission of the Aerospace Series, it combines elements of engineering and emerging technology with an accessible discussion of the important related legal and regulatory issues.

Peter Belobaba, Jonathan Cooper, and Allan Seabridge

Acknowledgements

I gratefully acknowledge the contributions of the following people: Michael Baum for listening to me and in some respects collaborating with me on a previous book; Peter La Franchi for providing valuable information from his perch in Australia; and of course, Sandi for her encouragement and support through the evolution of this book.

List of Acronyms and Abbreviations

3GPP	3rd Generation Partnership Project
AAM/UAS	Advanced Air Mobility/Urban Air Mobility
ACs	Advisory Circulars
ACAS	Aircraft Collision Avoidance System
ADR	Aerodrome
ADSB-Out	Automatic Dependent Surveillance Broadcast-Out
AFAC	Civil Aviation Federal Agency (Mexico)
AGL	Above Ground Level
AIAA	American Institute of Aeronautics and Astronautics
AIP	Aeronautical Information Publications
AIS	Aeronautical Information Service
ALJ	Administrative Law Judges
AMA	Academy of Model Aeronautics
AMC	Acceptable Means of Compliance or Alternate Means of Compliance
AMQP	Advanced Message Queuing Protocol
ANS	Air Navigation Services
ANSI	American National Standards Institute
ANSP	Air Navigation Service Providers
ARAG	Aviation Rulemaking Advisory Group (FAA)
ARC	Aviation Risk Category/Aviation Rulemaking Committee
ASBU	Aviation System Block Upgrade
ASTM	ASTM International (formerly American Society for Testing and Materials)
ATC	Air Traffic Control
ATCO	Air Traffic Control Organization
ATM	Air Traffic Management
ATO	Air Traffic Organizations
ATS	Air Traffic Services
AURA	Air Traffic Management U-space Project
BCAs	Bridge Certificate Authorities
BVLOS	Beyond Visual Line of Sight
C2	Command and Control
CA	Certificate Authority

CAA	Civil Aviation Authorities
CAAC	Civil Aviation Administration of China
CAAS	Civil Aviation Authority of Singapore
CANSO	Civil Air Navigation Services Organization
CAPSCA	Public Health events in civil aviation
CASA	Civil Aviation Safety Authority (Australia)
CAST	Commercial Aviation Safety Team
CIR	Commission Implementing Regulation
CIS	Common Information Service
CISA	Cybersecurity and Infrastructure Security Agency
CFR	Code of Federal Regulations
CNS	Communications/Command, Navigation, Surveillance
COA	Certificate of Authorization or Waiver (USA)
CONOPs	Concept of Operations
CORUS	Concept of Operation for EuRopean Unmanned Air Traffic Management Systems
CORUS-XUAM	Concept of Operations for European UTM Systems – Extension for Urban Air Mobility
CPDLC	Controller Pilot Data Link Communications
CSIS	Center for Strategic and International Studies
CU	Command Unit
C-UAS	Counter UAS
DAA	Detect and Avoid (system)
DAIM	Drone Aeronautical Information Management
DD/ASF	Deputy Director, Aviation Security and Facilitation
DDoS	Distributed Denial of Service
DNS	Domain Name System
DNSSEC	Domain Name System Security Extensions
DOC	Declaration of Compliance
DOD	Department of Defense (USA)
EASA	European Aviation Safety Agency
EC	European Commission
ENAC	Italian Civil Aviation Authority
ENCASIA	European Network of Civil Aviation Safety Investigation Authorities
EPAS	European Plan for Aviation Safety
EU	European Union
EUROCAE	European Organization for Civil Aviation Equipment
EUROCONTROL	European Organization for the Safety of Air Navigation
eVTOL	Electric-powered Vertical Takeoff and Landing
FAA	Federal Aviation Administration
FAR	Federal Aviation Regulations
FCC	Federal Communications Commission
FCC	Flight Control Computers
FCL	Flight Crew Licensing
FDIC	Federal Deposit Insurance Corporation

FIMS	Flight Information Management Systems
FLARM	Flight and Alarm
FMRA	FAA Modernization and Reform Act
FMS	Flight Management Systems
FTC	Federal Trade Commission
FUA/AFUA	Flexible Use of Airspace/Advanced Flexible Use of Airspace
GANP/ANP	Global and Regional Air Navigation Plan/Air Navigation Plan
GIS	Geographic Information Service
GM	Guidance Material
GNSS	Global Navigation Satellite System
GRPS	Global Risks Perception Survey
GSMA	Groupe Speciale Mobile Association
gTLD	generic Top-Level Domain
GUTMA	Global UTM Association
IAA	Irish Aviation Authority
IANA	Internet Assigned Numbers Authority
IATA	International Air Transport Association
IATF	International Aviation Trust Framework
ICC	Interstate Commerce Commission
ICAO	International Civil Aviation Organization
ICC	Interstate Commerce Commission
IEC	International Electrotechnical Commission
IEEE	Institute of Electrical and Electronics Engineers
IETF	Internet Engineering Task Force
IFR	Instrument Flight Rules
IIC	Investigator in Charge
ILT/ILENT	Environment and Transport Inspectorate
IoT	Internet of Things
IP	Internet Protocol
IPP	UAS Integration Pilot Program (USA)
IPv6	Internet Protocol version 6
IR	Implementing Rules
ISMS	Information Security Management System
ISO	International Organization for Standards
ITU	International Telecommunications Union
JAA	Joint Aviation Authorities
JARUS	Joint Authorities for Rulemaking on Unmanned Systems
JMS	Java Message Service
LAANC	Low Altitude Authorization and Notification Capability
LCOS	United Nations Convention on the Law of the Seas
LRC	Linear Responsibility Chart
LTE	Long Term Evolution
LUC	Light UAS Certificate
MLITT	Minister of Land, Infrastructure, Transport and Tourism (Japan)
MS	Member States

MTOW	Maximum Takeoff Weight
NAA	National Aviation Authority
NAS	National Airspace System
NASA	National Aeronautics and Space Administration (USA)
NGO	Non-governmental Organizations
NIST	National Institute of Science and Technology
NOA	Notice of Availability
NOTAM	Notice to Airmen
NPA	Notice of Proposed Amendment
NPRM	Notice of Proposed Rulemaking
NTSB	National Transportation Safety Board
OCHA	United Nations Office for the Coordination of Humanitarian Affairs
OPS	Operations
ORA	Operational Risk Assessment
PANS	Procedures for Air Navigations Services
PAO	Public Aircraft Operations
PDRA	Predefined Risk Assessment
PKI	Public Key Infrastructure
PSU	Providers of Services for UAM
QNH	Question Nil Height; Barometric Pressure Adjusted to Sea Level
RA	Registration Authority
RACI	Responsible, Accountable, Consulted, Informed
RID	Remote Identification
RFI	Requests for Information
RMT	Rulemaking Task
RPA	Remotely Piloted Aircraft
RPAS	Remotely Piloted Aircraft Systems
RPIC	Remote Pilot in Command
RTCA	Radio Technical Commission for Aeronautics
SACAA	South African CAA
SAE	Society of Automotive Engineers
SAFIR	Safe and flexible integration of initial U-space services in a real environment
SARPS	Standards and Recommended Practices
SDOs	Standards Development Organizations
SDSPs	Supplemental Data Service Providers
SEC	Securities and Exchange Commission
SERA	Standardized European Rules of the Air
SES	Single European Skies
SESAR-JU	Single European Sky Air Traffic Management Research Joint Undertaking
SIDs	Standard Instrument Departures
SIO	Systems Integration and Operationalization Demonstration Activity
SMM	The Safety Management Manual
SORA	Specific Operations Risk Assessment

SRM	Safety Risk Management
SSGC	Secretariat Study Group on Cybersecurity (ICAO)
SSR	Secondary Surveillance Radar
STARS	Standard Terminal Automation Replacement System
STS	Standard Scenario
sUAS	Small Unmanned Aircraft System
SUPPS	Regional Supplementary Procedure
SWIM	System Wide Information Management
TCL	Technical Capability Levels
TLD	Top Level Domain
TLS	Transport Layer Security
TMA	Terminal Maneuvering/Control Area
UA	Unmanned Aircraft
UAM	Urban Air Mobility
UAS	Unmanned Aircraft System
UAS-AG	Unmanned Aircraft Systems Advisory Group
UNCLOS	United Nations Conference on the Law of the Seas
UPP	UTM Pilot Program (USA)
USP	UTM Service Provider
USS	UAS Service Provider
UTM	Unmanned Aircraft Systems Traffic Management
UVR	UAS Volume Reservations
VFR	Visual Flight Rules
VHL	Very High Level
VLD	Very Large-Scale Demonstration
VLL	Very Low Level (airspace)
VLOS	Visual Line of Sight
VPN	Virtual Private Network
VUTURA	Validation of U-Space Tests in Urban and Rural Areas
WEF	World Economic Forum
WJHTC	William J. Hughes Technical Center (FAA)

1

Background

Introduction

Every civilized nation has some form of rules, regulations, policies, and laws that regulate economic activities and social behaviour. The scope and process for the creation of these guidelines varies widely among nations, depending upon many factors that derive from the form of government and cultures that produce those rules. As a society grows more open and more complex, more rules and regulations may become necessary to maintain order and protect people from harm that may result from the unrestricted activities of others. Those harms can be physical, health related, economic, environmental, or any number of potentially damaging outcomes from a governmental entity, an organization, or an individual doing or failing to do something that threatens the well-being of others. As the technology of unmanned aircraft systems has evolved and more potential users seek to deploy them for recreational, commercial, scientific, or public safety purposes, national and local governments have attempted to address the challenges presented by the integration of remotely piloted aircraft into national civil airspaces, particularly the perception of increased risk of harm to persons or property, by developing regulations that address the risk management and policy issues resulting from the use of these devices.

Since there is no settled overarching international law, treaty, or body of community-based standards that governs unmanned aerial systems (although several such efforts are underway), the developer, manufacturer, distributor, and end user must be wary of the potential for inadvertent violations of existing law, or of having a formerly permitted activity become illegal or proscribed as a result of a change in the rules. Methodologies for understanding the existing rules and participating in the process of developing new or modified rules should be key elements in the business plan for any individual or entity seeking to participate in the unmanned systems arena. Likewise, government regulators can benefit from comprehensive and adaptable criteria for the development of new technologies and the safe integration of those technologies into complex airspace systems.

Setting the Stage for Integration of Remotely Piloted Aircraft into Non-segregated Airspace

Like the oceans, the world's airspace is a public resource, owned by no one. Territorial waters adjacent to continents, islands, and land masses are controlled by the nations that claim sovereignty over the land contiguous to the oceans. International treaties such as the United Nations Conference on the Law of the Seas (UNCLOS) and United Nations Convention on the Law of the Seas (LCOS) establish the criteria for recognition and structure of national territorial waters, setting limits on the extent of those waters and economic zones. There are dozens of such treaties, conventions and instruments dealing with every conceivable aspect of human management of the world's oceans. Disputes often arise about the interpretation of treaty language and there is a forum to resolve those disputes. "Peaceful settlement of international disputes occupies an important place in international law in general and the law of the sea is no exception" (Tanaka 2012). The same can be said for the treaties and conventions that deal with international airspace, specifically the Convention on International Civil Aviation (also known as the Chicago Convention of 1944), which created the International Civil Aviation Organization (ICAO). The similarity between the legal regimes of international and domestic airspace and the high seas (and coastal waters) bears examination, for many aviation regulations are derived from ancient maritime common law.

The Law of the Sea and the Law of the Air

Perhaps the greatest challenge to lawmakers and regulators charged with maintaining an acceptable level of safety in their respective national airspace systems is how to integrate unmanned aircraft into existing aeronautical environments where manned aircraft have been navigating with ever increasing levels of safety for decades. The law of the sea evolved over centuries of commercial shipping activities, and is one of the oldest branches of public international law (Tanaka 2012). The law of the airways has a relatively shorter life, but it has progressed and expanded at a far greater pace than the law of the sea. What makes the laws of the sea and the air similar enough to warrant comparison, and perhaps guidance for aviation professionals and regulators, is that both bodies of law, policy, and regulation deal with access to and the safety of a vast common community resource, to wit: the oceans and the airspace above the Earth and the high seas. While it seems unlikely that two vessels navigating the vastness of the world's oceans could ever collide, they do, far more often than may commonly be known. The European Maritime Safety Agency Annual Overview of Marine Casualties and Incidents 2018 reported an average of 3315 occurrences per year for the four years ending in 2017. In 2017 more than 1500 cargo ships were involved in accidents that resulted in 25 fatalities. Casualties numbered 1018 persons, half related to issues of a navigational nature, such as contacts, grounding/stranding, and collisions. The same observation could be made about the skies. The skies are immense, seemingly limitless, yet airplanes collide, often with catastrophic results. Thus, in both realms, the sea and the sky, regulations and "rules of the road" are necessary to minimize the risk of such events. The aviation rules for airplanes occupying the same airspace in close proximity to one another generally follow the rules at sea (for example, "... when aircraft of the same category are converging at approximately the same altitude [except head-on, or nearly so], the aircraft to the other's right has the right-of-way"; similarly, "... when two power-driven vessels are meeting on reciprocal or nearly reciprocal courses so as to involve risk of collision each shall alter her course to starboard so that each shall pass on the port side of the other").

A Brief History of Aviation Regulations

Although the first powered flight with a pilot on board is attributed to the Wright Brothers (specifically with Orville, the daring one, at the controls) at 10:35 a.m. on 17 December 1903 near Kitty Hawk, North Carolina, the first aviation regulations did not appear for another eight years when Great Britain passed the Aerial Navigation Act of 1911. The British followed that law with a second Aerial Navigation Act of 1913, which transferred the control of aviation activities to the Secretary of State for War. In 1916 the British Air Board was created, which greatly influenced the post-World War I control of civil aviation in Great Britain (Chaplin 2011).

Although the United States did not initiate the first efforts to create aviation safety regulations, the US system of regulations, whether aviation related or not, bears closer examination because the US is historically viewed as a leader in the evolution of aviation regulations world-wide. Regulations, and the supporting standards and guidance materials that help people and organizations comply with them, have been a core component of the legal system in the United States for nearly 150 years. Every federal regulatory agency promulgates rules that are intended to carry out the agency's legislative mandates. State and local regulatory agencies or commissions act in much the same way.

Lawmakers in most countries generally do not have the luxury of time, expertise, or the resources to define and monitor every element of the particular industry or activity that they undertake to regulate through the enactment of laws that are national in scope. The agencies that the US Congress creates and funds, for example, are delegated the legislative powers that Congress has been granted by Article, Section 1 of the US Constitution. Those delegated powers are implemented through the rule-making process. The oldest federal regulatory agency that still exists in the US is the Office of the Comptroller of the Currency, which was created in 1863 to charter and regulate the nation's banks. The modern era of federal regulation really began with the creation of the Interstate Commerce Commission (ICC) in 1887, which was directed to protect the public from excessive and discriminatory railroad rates. The regulation was economic in nature, by the setting of rates and by regulating how railroad services were to be provided. The ICC's administrative model was that of an independent, bipartisan commission that used an adjudicatory approach to arrive at decisions on contested matters. This structure was adopted by several subsequently created agencies, such as the Federal Trade Commission (FTC), the Water Power Commission (later the Federal Power Commission), and the Federal Radio Commission (which became the Federal Communications Commission). Congress created several other agencies early in the 20th century to regulate commercial and financial systems – including the Federal Reserve Board, the Tariff Commission, the Packers and Stockyards Administration, and the Commodities Exchange Authority, all established before 1922. The Food and Drug Administration was created in 1931 to ensure that certain foods and drugs were fit for human consumption.

Many other federal regulatory agencies were created in the 1930s as part of President Franklyn D. Roosevelt's "New Deal" programs. These included the Federal Home Loan Bank Board (1932), the Federal Deposit Insurance Corporation (FDIC) (1933), the Commodity Credit Corporation (1933), the Farm Credit Administration (1933), the Securities and Exchange Commission (SEC) (1934), and the National Labor Relations Board (1935). The jurisdictions of both the Federal Communications Commission (FCC) and the Interstate Commerce Commission were also expanded to regulate other methods of communications (e.g. telephone and telegraph) as well as other transportation modes (trucking, pipelines, and common carriers such as bus lines and telephones). The ICC was dissolved in 1995, and its remaining powers were reassigned to the Surface Transportation Board.

The United States' first attempt to regulate commercial aviation arose out of the US Post Office's commencement of airmail operations in 1918. The first published aviation regulations were released in 1926 by the Aeronautics Branch of the Department of Commerce, and were known as "Air Commerce Regulations." They consisted of six chapters spanning 45 pages of text, and covered the areas of Licensing, Marking, and Operation of Aircraft, Licensing of Pilots and Mechanics, Air Traffic Rules, and were rounded out by a miscellaneous section. Aviation is indeed one of the oldest regulated industries or activities in the US. Presently the Federal Aviation Regulations (FARs) fill four volumes of the Code of Federal Regulations, consisting of over 460 sections filling more than 3600 pages, and totaling around 8400 regulations, counting major subparts and sections.

International Civil Aviation Regulations

Outside of the US, aviation regulations vary from country to country, as can be expected. Some countries have aviation rules and regulations similar to the US, but few, if any, are as comprehensive and voluminous. In Europe, for example, each individual country (Member State) has its own set of regulations, developed and enforced by their domestic civil aviation authorities (CAAs). An additional layer of safety rules falls under the regulatory and administrative jurisdiction of the European Aviation Safety Agency (EASA), an agency of the European Union (EU), established in 2002 and consisting of 31 Member States (27 European Union states plus Switzerland, Norway, Iceland, and Liechtenstein observer states). EASA's role is to provide advice to the European Union for drafting new legislation, implementing and monitoring safety rules, including inspections in the Member States, type certification of aircraft and components, as well as the approval of organizations involved in the design, manufacture, and maintenance of aeronautical products, authorization of third-country (non-EU) operators, safety analysis, and research.

The European Organization for the Safety of Air Navigation, known as "EUROCONTROL," is the entity that provides harmonized air navigation services across European skies, and is separate from EASA. In 2011 the European Union established another organization, the European Network of Civil Aviation Safety Investigation Authorities (ENCASIA), via Regulation No. 996/2010, whose "... mission is to further improve the quality of air safety investigations and to strengthen the independence of the national investigating authorities." This organization makes safety recommendations to EASA, but does not create regulations. Thus, each member state has the option to create and maintain its own aviation safety organization with attendant powers to investigate accidents and impose sanctions on violators of their domestic regulations.

The 1944 Chicago Convention on International Civil Aviation (which created the International Civil Aviation Organization, ICAO) generally and comprehensively sets forth international standards and recommended practices for aviation (as provided in Article 37). Consistent with Article 26 of the Convention, Annex 13 deals with aircraft accident and incident investigation. The Annex states that investigation of serious accidents and incidents is to be conducted by the State where the accident or incident occurs, or where the State of Registry when the location of the accident or serious incident cannot definitely be established as being in the territory of any State. In addition, a State may delegate the task of conducting the investigation to another State or request its assistance. By virtue of this Annex, EASA should be invited to participate in a safety investigation "... in order to contribute, within the scope of its competence, to its efficiency and to ensure the safety of aircraft design, without affecting the independent status of the investigation. National civil aviation authorities should be similarly invited to participate in safety investigations."

The foregoing structure is similar to the United States' system, where all aviation regulations and air navigation services fall under the purview of one agency, the Federal Aviation Administration (FAA), a subdivision of the US Department of Transportation (DOT). However, investigations of aviation accidents are solely vested in another agency, the National Transportation Safety Board (NTSB), an independent agency of the United States Government, and, unlike the FAA, is not a sub-agency of the Department of Transportation. The NTSB has no regulatory or enforcement powers. In addition to investigating transportation accidents, the NTSB also provides administrative, quasi-judicial review process for FAA enforcement actions by Administrative Law Judges (ALJs), who preside over a wide variety of FAA enforcement matters. In summary, the major difference between the European and American systems is that each EU Member State creates its own aviation regulations, and commits to harmonizing their domestic regulations with EASA and EU aviation safety standards, as well as remaining in compliance with the International Civil Aviation Organization's requirements. In contrast, there is only one overarching set of aviation regulations in the US, and the individual states generally have little or no power to create their own regulations. The legal and policy tug-of-war between state, local, and federal government regulatory power over aviation activities is controversial, falling under the broad theme of "preemption," but that is a discussion for another book. Some state and local governments have enacted legislation that impact low-level (generally below 400 ft AGL) unmanned aircraft operations, and the extent to which they can do so without running afoul of federal law, and the FAA's declared policy, has yet to be definitively decided in US courts. There is no such gray area between EU Member State regulations and the rules of the road established and agreed upon by the members of EUROCONTROL and EASA.

Another major distinction between EASA and NTSB is that the former is only chartered to oversee aviation safety, whereas the NTSB also has more or less exclusive jurisdiction over other non-aviation transportation sectors, such as railway, highway, marine, and pipeline. NTSB investigators will respond anywhere in the world to aviation accidents involving products manufactured in the US. The NTSB "Go Teams" only respond to accidents that occur on US territory or in international waters. Elsewhere, the lead investigator is by default the government in whose territory the accident occurs, often assisted by a US "accredited representative" from the NTSB's staff of "investigators in charge" (IICs) if a US carrier or US manufactured airplane is involved.

Virtually every member of the United Nations (currently numbering 193 Member States and 2 non-member observer states) has its own version of domestic aviation regulations, which for the most part is patterned after the Chicago Convention articles and annexes. As noted above, the Chicago Convention established the International Civil Aviation Organization as a means to secure international cooperation and the highest possible degree of uniformity in regulations and standards, procedures, and organization regarding civil aviation matters. The Convention produced the foundation for a set of rules and regulations regarding air navigation as a whole, with the intent to enhance safety in flying by laying the groundwork for the application of a common air navigation system throughout the world.

ICAO works in close cooperation with other members of the United Nations family, such as the World Meteorological Organization, the International Telecommunication Union, the Universal Postal Union, the World Health Organization, the International Maritime Organization and the Arctic Council. Aviation-related non-governmental organizations (NGOs) also participating in ICAO's efforts include the International Air Transport Association, the Airports Council International, the International Federation of Air Line Pilots' Associations, and the International Council of Aircraft Owner and Pilot Associations.

A comprehensive analysis of the aviation regulatory schemes in each and every country that publishes some form of regulation is beyond the scope of this book. However, since ICAO Member States contract to follow ICAO's rules and supplementary material, and to publish for all to see any exceptions taken to any ICAO Article, Annex, Regional Supplementary Procedure ("SUPPS") or Procedures for Air Navigations Services (PANS), the beginning point must be the aforementioned ICAO rules and procedures themselves, with particular focus on those provisions related to safe integration and management of domestic and international airspace.

The Chicago Convention and the International Civil Aviation Organization

The Aeronautical Commission of the Peace Conference of 1919 (otherwise known as the Versailles Treaty) created an international agreement (the Convention for the Regulation of Aerial Navigation) that recognized that the airspace above the high seas was not as "free" as the oceans beneath. The contracting States to that Convention agreed that the States had exclusive jurisdiction over the airspace above the land and territorial waters of the States, but agreed to allow, in times of peace, innocent passage of civil aircraft of other States through their sovereign airspace, so long as the other provisions of the Convention were observed. States still retained the right to create prohibited areas in the interests of military needs or national security. During the years leading up to World War II and throughout that conflict the United States initiated studies and later consulted with its major allies regarding further harmonization of the rules of international airspace, building upon the 1919 Convention. Anticipating the pending termination of the hostilities and desiring to re-establish international transport by air, the US government invited 55 states and civil aviation authorities to attend a meeting to discuss these issues and to promote cooperation and "create and preserve friendship and understanding among the nations and peoples of the world," and in November 1944, an International Civil Aviation Conference was held in Chicago. Fifty-four nations (they are referred to as "States" in the Convention) attended this conference, and 52 of those nations signed the new *Convention on International Civil Aviation*. The convention created a specialized agency, the International Civil Aviation Organization (ICAO), to oversee the terms of the Convention, and as a means to secure international cooperation and the highest possible degree of uniformity in regulations and standards, procedures, and organization regarding civil aviation matters. The Chicago Conference laid the foundation for a set of rules and regulations regarding air navigation as a whole, which was intended to enhance safety in flying and construct the groundwork for the application of a common air navigation system throughout the world.

ICAO's many objectives are set forth in the 96 Articles of the Chicago Convention and the 18 annexes thereto. Numerous published supplements (Standards and Recommended Practices, or SARPS) and Procedures for Air Navigation Services (PANS) (which are under continual review and revision) set forth additional standards and guidelines for Contracting States. These Contracting States may take exception to any element of the annexes, and those exceptions are also published. Contracting States are also responsible for developing their own aeronautical information publications (AIPs), which provide more detailed information to ICAO and other States about air traffic, airspace, airports, navaids (navigational aids), special use of airspace, weather, and other relevant data that are available for use by aircrews arriving into or transiting through the State's airspace. The AIPs also contain information about the State's exceptions to the annexes and any significant differences between the rules and regulations of the State and ICAO's rules.

The annexes cover personnel licensing (Annex 1), rules of the air (Annex 2), meteorological services for international air navigation (Annex 3), aeronautical charts (Annex 4), units of measurement to be used in air and ground operations (Annex 5), operation of aircraft (Annex 6), aircraft nationality and registration marks (Annex 7), airworthiness of aircraft (Annex 8), facilitation of border crossing (Annex 9), aeronautical communications (Annex 10), air traffic services (Annex 11), search and rescue (Annex 12), aircraft accident investigation (Annex 13), aerodromes (Annex 14), aeronautical information services (Annex 15), environmental protection (Annex 16), security-safeguarding international civil aviation against acts of unlawful interference (Annex 17), and the safe transportation of dangerous goods by air (Annex 18). The only reference to unmanned aircraft in the Convention is in Article 8, which states that:

No aircraft capable of being flown without a pilot shall be flown without a pilot over the territory of a contracting State without special authorization by that State and in accordance with the terms of such authorization. Each contracting State undertakes to ensure that the flight of such aircraft without a pilot in regions open to civil aircraft shall be so controlled as to obviate danger to civil aircraft.

ICAO's rules apply to international airspace, which is typically defined as the airspace over the high seas more than 12 miles from the sovereign territory of a State (country), as well as some domestic airspace by virtue of incorporation into a contracting State's own regulatory scheme. The rules just apply to the 193 Member States, so any nation that declines to become an ICAO member is not entitled to the protection of ICAO's rules. However, ICAO is a voluntary organization and there are no provisions for enforcement of the regulations or standards such as those found in the US Federal Aviation Regulations or similar regulatory provisions enacted by the Member States. As a practical matter, a nation that elects to not be an ICAO Member State will find that no foreign air carriers will be allowed to land in that nation's airports, except in an emergency, and that nation's airlines, if any, will not be permitted to operate in the airspace of a Member Nation.

However, even a Member State may find itself isolated and ostracized for a blatant violation of ICAO rules, such as what happened to Belarus in May of 2021 after its leadership engineered the diversion and forced landing of a commercial airline's aircraft in Belarus to detain a political dissident journalist (basically a hijack). The airline (Ryanair) was operating a scheduled flight on an approved international route over Belarus, not intended to land in that country. Belarus military aircraft forced the airliner to land in Belarus and the dissident was taken off the aircraft and placed into custody. In response, the European Union banned Belavia, the national airline of Belarus, from EU airspace, effectively shutting the airline down indefinitely. Other EU Member States and non-EU nations also ceased flights in and out of Belarus, and the US Department of Transportation (at the request of the State Department) issued a final order blocking most travel between the United States and Belarus. All EU nations are ICAO Member States. Articles 6 of the Convention gives Contracting States the authority to revoke permission of another State's aircraft to operate over the territory of the Contracting State, so this action by the EU was wholly consistent with the Convention, as well as ICAO's rule.

As a founding member of ICAO and a nation that has a substantial interest in preserving harmony in international commercial aviation, the United States enforces ICAO's rules against US operators to the extent that the ICAO rule has been incorporated into the FARs and does not conflict with domestic regulations. Other ICAO Member States do the same. In 2012 ICAO created a Remotely Piloted Aircraft Systems (RPAS) Panel, designed to deliver standards for unmanned aircraft to ICAO's governing council in 2018. The panel's focus was and is on the development of standards and recommended practices (SARPs) for adoption by the ICAO Council regarding airworthiness, operations, and licensing of remote pilots. The first edition of the "Manual on Remotely Piloted Aircraft Systems" was released in 2015. Also in 2015, ICAO established the Unmanned Aircraft Systems Advisory Group (UAS-AG) to support the Secretariat in developing guidance material and expedite the development of provisions to be used by Member States to regulate unmanned aircraft systems. The Advisory Group has continued to respond to ICAO's Requests for Information (RFI) in 2017, 2018, and 2019 by holding industry symposia to showcase submissions by interested parties responding to the RFIs. The last scheduled symposium, "Drone Enable/2021" was held virtually in April 2021. The 2020 Rio de Janeiro Symposium was canceled due to the Covid-19 pandemic.

The following chapter will offer a brief look at regulatory efforts around the world that focus on integration of drones into domestic airspace. The European Union is developing its version of unmanned aircraft systems traffic management, called "U-space," and the US is engaged in a parallel effort labeled "UTM," while also participating in the European effort with the intent to harmonize these two major projects. Most other non-EU nations have thus far not taken the deep dive into the integration technology that is ongoing in the US and Europe, but those two undertakings will presumably provide a template for other nations to adopt. The overall picture of global airspace integration efforts intended to merge UAS operations with manned aviation at low levels, as well as higher controlled and uncontrolled airspace, divides the strategies into three broad categories: Unmanned Aircraft Systems Traffic Management (UTM) in the US, "U-space" in the European Union, and an FIMS-based (Flight Information Management Systems) data exchange gateway that connects UTM participants with a nation's air traffic management system, Australia being a good example. Air traffic management (ATM) is the broad descriptor of air traffic control services in all categories airspace.

The details of those strategies are discussed in Chapters 2, 3, and 4.

Conclusion

Aviation regulations have been in place for over a century. They have evolved into a complex and interconnected web of overlapping regulations, acceptable means of compliance, guidance materials, and standards. The introduction of unmanned aircraft into national airspaces around the world may properly be characterized as the greatest technical innovation in aviation since the development and implementation of gas turbine engines for commercial aircraft. This phenomenon has also been the most disruptive to airspace safety, something that cannot be said for jet engines. The next chapter takes us to UAS airspace integration efforts in the European Union, which have been responsive to this disruptive technology.

References

14 Code of Federal Regulations §91.113(d)(Converging).

Chaplin, J.C. (2011). Safety Regulation – The First 100 Years, *Journal of Aeronautical History*, Paper No. 2011/3.

https://ec.europa.eu/transport/modes/air/encasia_en.

https://www.ntsb.gov/investigations/process/Pages/default.aspx.

2

UAS Airspace Integration in the European Union

EASA leads the way in Europe in developing a roadmap for integration of remotely piloted aircraft into the 31 Member States' domestic airspace, seeking uniformity and harmonization across and between national borders. The agency has published a series of documents addressing remotely piloted aircraft in EU airspace. They set forth guidelines (GM), acceptable means of compliance (AMC), and regulations for the operation of UAS in Member States' domestic airspace. The documents cited herein are under continual review and amendment, so the prudent reader should confirm the last applicable revision or amendment by accessing the EASA website at: https://www.easa.europa.eu/home.

Regulations, Opinions, Decisions

Regulation (EU) 2018/1139 of the European Parliament

The full title of this regulations is: Regulation (EU) 2018/1139 of the European Parliament and of the Council on common rules in the field of civil aviation and establishing a European Union Aviation Safety Agency, and amending Regulations (EC) No. 2111/2005, (EC) No. 1008/2008, (EU) No. 996/2010, (EU) No. 376/2014 and Directives 2014/30/EU and 2014/53/EU of the European Parliament and of the Council, and repealing Regulations (EC) No. 552/2004 and (EC) No. 216/2008 of the European Parliament and of the Council and Council Regulation (EEC) No. 3922/91.

This regulation is commonly known as the "Basic Regulation."

Articles 2(1) and 55 through 57 deal with unmanned aircraft.

Articles 2(1)(a) and (b) define the scope of the regulation, namely the design and production of products, parts, and equipment to control aircraft remotely, and are subject to registration in Member and non-member states that they are to operate in, reside in, or have its principal place of business within the territory of a Member State.

Article 55 (Essential requirements for unmanned aircraft) states: "The design, production, maintenance and operation of aircraft referred to in point (a) and (b) of Article 2(1), where it concerns unmanned aircraft, and their engines, propellers, parts, non-installed equipment and equipment to control them remotely, as well as the personnel, including

UAS Integration into Civil Airspace: Policy, Regulations and Strategy, First Edition. Douglas M. Marshall.
© 2022 John Wiley & Sons Ltd. Published 2022 by John Wiley & Sons Ltd.

remote pilots, and organizations involved in those activities, shall comply with the essential requirements set out in Annex IX, and, where the delegated acts referred to in Article 58 and the implementing acts referred to in Article 57 so provide, with the essential requirements set out in Annexes II, IV and V."

Article 56 (Compliance of unmanned aircraft) lists seven requirements for design, production, and operation of unmanned aircraft. The physical and operational characteristics of unmanned aircraft and the areas where they operate *may* require a certificate for the aircraft and for the pilot and operator. A certificate *shall* be issued when the applicant demonstrates compliance with delegated acts set forth in Articles 57 and 58. The certificate *shall* specify the safety-related limitations, operating conditions and privileges. The certificate *may* be amended to add or remove limitations. The certificate *may* be suspended or revoked for non-compliance with conditions, rules, and procedures required to maintain the certificate. A declaration of compliance with those limitations, conditions, and privileges *may* be required. Under certain listed circumstances, those essential requirements and those detailed rules *shall* constitute "Community harmonization legislation" within the meaning of Regulation (EC) No. 765/2008 of the European Parliament and of the Council and Decision No. 768/2008/EC of the European Parliament and of the Council. Member States *shall* ensure that information about registration of unmanned aircraft and of operators of unmanned aircraft that are subject to a registration requirement is stored in digital, harmonized, interoperable national registration systems. Member States *shall* be able to access and exchange that information through the repository referred to in Article 74. Last, this Section *shall* be without prejudice to the possibility for Member States to lay down national rules to make subject to certain conditions the operations of unmanned aircraft for reasons falling outside the scope of this Regulation, including public security or protection of privacy and personal data in accordance with the Union law (emphasis added).

Article 57 (Implementing acts as regards to unmanned aircraft), the Commission *shall* adopt implementing acts laying down detailed provisions concerning:

(a) the specific rules and procedures for the operation of unmanned aircraft as well as for the personnel, including remote pilots, and organizations involved in those operations;

(b) the rules and procedures for issuing, maintaining, amending, limiting, suspending, or revoking the certificates, or for making declarations, for the operation of unmanned aircraft as well as for personnel, including remote pilots, and organizations involved in those activities, and for the situations in which such certificates or declarations are to be required; the rules and procedures for issuing those certificates and for making those declarations may be based on, or consist of, the detailed requirements referred to in Sections I, II, and III;

(c) the privileges and responsibilities of the holders of certificates and of natural and legal persons making declarations;

(d) the rules and procedures for the registration and marking of unmanned aircraft and for the registration of operators of unmanned aircraft, referred to in Section 4 of Annex IX;

(e) the rules and procedures for establishing digital, interoperable, harmonized, national registration systems referred to in Article 56(7); and

(f) the rules and procedures for the conversion of national certificates into the certificates required under Article 56(1).

Article 58 (Delegated powers) describes the Commission's powers to lay down detailed rules regarding the requirements, privileges, and responsibilities set forth in Articles 55 through 57.

Annex IX sets forth the essential requirements for the design, production, maintenance, and operation of unmanned aircraft, for environmental performance, and for registration and marking of unmanned aircraft and their operators.

All of these requirements are intended to facilitate the introduction and integration of unmanned aircraft systems into the national airspaces of EU Member States. Refer to Rulemaking Tasks (RMTs) set forth in the **EASA European Plan for Aviation Safety 2021–2025 Volume II**, discussed below, for more detail.

Commission Implementing Regulation (EU) 2019/947

The Implementing Regulation provides a framework for the safe operation of drones in European skies (which includes both EU and EASA Member States). The rules adopt a risk-based approach to regulation, and do not distinguish between commercial drone operations and hobbyists. The primary categories for airspace access are the weight and physical specifications of the drone, and the scope of the intended operation. CIR 2019/947 went into effect on 30 December 2020. The regulations created three categories of operations: "open," "specific," and "certified."

The "open" category covers lower risk operations, where safety is ensured by drone operator compliance with the relevant regulatory requirements for the intended operation. This category is subdivided into three subcategories labeled "A1," "A2," and "A3." No prior authorization is required before starting a flight because operational risks in this category are considered to be low.

The "specific" category includes higher risk operations, where the drone operator obtaining an operational authorization from the national competent authority before starting the operation ensures safety. The drone operator is required to conduct a safety risk assessment before seeking authorization, which will establish the requirements necessary for safe operations.

In the "certified" category, the safety risk is substantially higher than the other two categories, so that certification of the drone operator and the aircraft by the national competent authority is required to ensure safety. In addition, the licensing of the remote pilot(s) is mandatory.

The management of air traffic and air traffic services for drone integration with manned aviation will be ensured through the U-space. U-space is another arm of the drones' regulatory framework. It creates and harmonizes the necessary conditions for manned and unmanned aircraft to operate safely in the U-space airspace, so as to prevent collisions between aircraft and to mitigate the air and ground risks. The U-space regulatory framework, supported by clear and simple rules, should permit safe aircraft operations in all areas and for all types of unmanned operations. This is the airspace architecture and the services that will ensure the safe operation of drones once in flight. The regulatory framework for U-space was published in the Official Journal of the European Union on 22 April 2021 and is discussed below.

Commission Delegated Regulation 2019/945

This Regulation sets forth the requirements for the design and manufacture of unmanned aircraft systems that are intended to operate under the rules and conditions defined in the Implementing Regulation 2019/947, including remote identification (RID) add-ons (as set out in Part 6 of the Annex to this Regulation). The types of UAS, whose design, production, and maintenance are subject to certification, are defined.

The Regulation establishes rules on making a UAS intended for use in the "open" (low-risk) category and remote ID add-ons available in the European market, and on their free movement within the Union. Also included are the rules for third-country (non-EU) UAS operators conducting operations within the single European sky airspace pursuant to iImplementing Regulation 2019/947.

The full scope of the Regulation excludes privately built UASs (basically hobbyists and recreational flyers) that display a class identification label indicating which of the five UAS classes defined in this Regulation it belongs to. Chapter II of the Regulation applies to a UAS operated under the rules applicable to the "certified" and "specific" categories as defined therein. Chapter IV deals with UAS operators who have their principal place of business or reside in a third country, if the UAS are operated in the Union.

This Regulation does not apply to UASs intended exclusively for operations indoors.

Executive Director Decision 2019/021/R Issuing AMC and GM to CIR (EU) No. 2019/947

Article 1 states: "The Acceptable Means of Compliance and Guidance Material to the Annex to Commission Implementing Regulation (EU) 2019/947 are those laid down in Annexes I and II to this Decision." Those Annexes set forth the rules and procedures for the operation of unmanned aircraft, and follow the mandates of Article 76(3) of Regulation (EU) 2018/1139 to issue specifications and acceptable means of compliance and guidance material for the application of that Regulation. The Decision verifies that acceptable means of compliance and guidance materials are non-binding standards that may be used by persons and organizations to demonstrate compliance with the requirements of Regulation (EU) 2018/1139. EASA is mandated to reflect the state of the art and best practices in the field of aviation, and AMCs and GMs are needed to facilitate compliance with the regulations through harmonized implementation of CIR (EU) 2019/947.

The Decision also acknowledges that the methodology developed by the Joint Authorities for Rulemaking on Unmanned Systems (JARUS), known as "SORA" (specific operation risk assessment), as well as a predefined risk assessment (PDRA) for operations beyond visual line of sight (BVLOS) include internal and external consultation processes similar to those defined in EASA's rule-making procedures.

Explanatory Note to ED Decision 2019/021/R

The objective of the Decision is to maintain a high level of safety for UAS operations in the "open" and "specific" categories, which are expected to improve the harmonization of unmanned aircraft within the European Union. This is accomplished by the introduction of a regulatory framework for rules and procedures for the operation of unmanned aircraft and the first issue of AMCs and GMs related to Commission Implementing Regulation (EU) 2019/947. The Note explains why a new AMC and GM was needed (mainly focused on the "specific" category of UAS operations), how they were developed using the JARUS SORA process, what the objectives of the effort consisted of, and how EASA intends to achieve those objectives.

Specifically, the Decision describes in detail the goals of implementing an operation-centric, proportionate, risk- and performance-based regulatory framework for all UAS operations in the "open" and "specific" categories; ensuring a high and uniform level of safety for UAS; fostering the development of the UAS market; and addressing citizens' concerns about safety, privacy, data protection, and environmental protection.

Easy Access Rules for Unmanned Aircraft Systems (Regulations (EU) 2019/947 and (EU) 2019/945)

The most recent version of this 309-page document (released 30 September 2021), which is intended to "provide stakeholders with an updated, consolidated, and easy-to-read publication," is "not an official publication of EASA." The content is arranged by collecting the relevant officially published regulations, along with related Acceptable Means of Compliance and Guidance Materials that have been adopted as of January 2021. The sections are color coded to be user friendly: blue for implementing or delegated rule annexes; yellow for acceptable means of compliance; and green for guidance material.

Incorporated amendments reference Implementing Rules/Commission Regulations (EU) 2019/947, 2020/639, 2020/746, and 2021/1166; Delegated Rules/Commission Regulations (EU) 2019/945 and 2020/1058; and AMC & GM to IRs/ED decisions 2019/021/R and 2020/022/R.

The Cover Regulation to Implementing Regulation (EU) 2019/947, adopted May 24, 2019, delineates 29 specific areas of regulatory concerns for operation of unmanned aircraft in the European Union Member States. Number 26 deals with U-space, and is the only section directly devoted to airspace integration, although the term "U-space" is used repeatedly throughout the document to identify U-space as a potential conflict avoidance strategy. The section states:

> While the "U-space" system including the infrastructure, services and procedures to guarantee safe UAS operations and supporting their integration into the aviation system is in development, this Regulation should already include requirements for the implementation of three foundations of the U-space system, namely registration, geo-awareness and remote identification, which will need to be further completed.

The document republishes all 23 Articles set forth in CIR (EU) 2019/947, plus an Annex with three subparts (Part A, UAS Operations in the "open" category; Part B, UAS Operations in the "specific" category; and Part C, Light UAS Operator Certificate). The same format follows for the 42 Articles in CIR (EU) 2019/945, plus the Annex and its 17 subparts.

The document then provides detailed Guidance Material and Acceptable Means of Compliance for each Article and many subparts. It is intended to clarify the language of these regulations and offer ways to comply with them, and, where necessary, how to interpret the requirements or understand the underlying reasoning behind them. For example, CIR (EU) 2019/947 Article 2, Definitions, declares in Article 2(3) that: "'assemblies of people' means gatherings where persons are unable to move away due to the density of the people present."

These "Easy Access Rules" offer the following by way of explanation in "GM Article 2(3) Definitions:"

Definition of "Assemblies of People"

Assemblies of people have been defined by an objective criterion related to the possibility for an individual to move around in order to limit the consequences of an out-of-control UA. It was indeed difficult to propose a number of people above which this group of people would turn into an assembly of people: numbers were indeed proposed, but they showed quite a large variation. Qualitative examples of assemblies of people are:

(a) sport, cultural, religious, or political events;
(b) beaches or parks on a sunny day;
(c) commercial streets during the opening hours of the shops; and
(d) ski resorts/tracks/lanes.

Similarly, Article 2(11) "Definitions" of the Regulation declares:

"dangerous goods means articles or substances, which are capable of posing a hazard to health, safety, property or the environment in the case of an incident or accident, that the unmanned aircraft is carrying as its payload, including in particular:

(a) explosives (mass explosion hazard, blast projection hazard, minor blast hazard, major fire hazard, blasting agents, extremely insensitive explosives);

(b) gases (flammable gas, non-flammable gas, poisonous gas, oxygen, inhalation hazard);

(c) flammable liquids (flammable liquids; combustible, fuel oil, gasoline);

(d) flammable solids (flammable solids, spontaneously combustible solids, dangerous when wet);

(e) oxidizing agents and organic peroxides;

(f) toxic and infectious substances (poison, biohazard);

(g) radioactive substances;

(h) corrosive substances."

The Easy Access Rules for that Article state:

Definition of "Dangerous Good"

Under the definition of dangerous goods, blood may be considered to be capable of posing a hazard to health when it is contaminated or unchecked (potentially contaminated). In consideration of Article 5(1)(b)(iii):

(a) medical samples such as uncontaminated blood can be transported in the "open", "specific," or "certified" categories;

(b) unchecked or contaminated blood must be transported in the "specific" or the "certified" categories. If the transport may result in a high risk for third parties, the UAS operation belongs to the "certified" category (see Article 6 1 (b) (iii) of the UAS Regulation). If the blood is enclosed in a container such that in case of an accident, the blood will not be spilled, the UAS operation may belong to the "specific" category, if there are no other causes of high risk for third parties.

The risk assessment and mitigation sections of the two covered regulations cannot be overlooked. This document outlines a comprehensive SORA-based approach to determining air risk categories ("ARCs"), which will be a core component of the U-space system. Notionally, the U-space service providers will be tasked with providing the analysis of the air and ground risk autonomously and in four-dimensional space. It is the responsibility of the UAS operator to ensure that their detect and avoid tactical mitigation strategies satisfy the safety requirements of the desired airspace. These could include on-board systems such as ADS-B out, "FLARM" (flight and alarm), or operational procedures. U-space data service providers could also provide an air collision risk map (static or dynamic) that would inform an application of strategic mitigations to determine the initial air risk class.

Section C.3.2, "SORA U-space assumptions," notes that the SORA process "has used U-space mitigations in a limited extent because U-space is in the early stages of development. When U-space provides adequate mitigation to limit the risk of UAS encounters with manned aircraft, a UAS operator can apply for, and obtain, credit for these mitigations, whether they are tactical or strategic."[1]

Section C.5.2: "In the future, as U-space structures and rules become more readily defined and adopted, they will provide a source for the strategic mitigation of UAS operations by common structure and rules that UAS operators could more easily apply." (This should be contrasted with a "procedural separation system" as a prelude to U-space-C.5.2.2. This is the transition mitigation before U-space, and is a step in the evolution of airspace integration rules and regulations.) (See SERA 2021 for 1/26/23 implementation date of requirements for manned aircraft in U-space.)

Section C.6.3: EASA expects that UAS operations will make the VLL airspace more crowded, requiring more common structures and rules to lower the collision risk. They anticipate that U-space services will provide these risk mitigation measures.

UTM/U-space is repeatedly referenced in the Easy Access document, but only as a possible future application of an automated traffic management tool that is under development, acknowledging that these systems do not yet exist (as of the writing of the document). The goal of the technology is to keep the unmanned aircraft within its operational volume, as defined by GNSS, satellite systems, air traffic management, or U-space.

---

1 "Specific Operations Risk Assessment," a process to create, evaluate, and conduct an Unmanned Aircraft System operation. Available at: http://jarusrpas.org/sites/jarus-rpas.org/files/jar_doc_06_jarus_sora_v2.0.pdf.

The European Commission has issued three Implementing Regulations that have a direct impact on airspace integration efforts in European skies, with specific focus on remotely piloted aircraft and U-space airspace. These regulations anticipate that RPAs will operate in some designated airspaces along with manned aircraft, and generally require that manned aircraft entering those designated airspaces will be required to make themselves conspicuous to the U-space service providers and UAS operators so as to avoid conflicts and hazardous proximity to one another.

Annex 1, Issue 1 to ED Decision 2019/021/R "Acceptable Means of Compliance (AMC) and Guidance Material (GM) to Commission Implementing Regulation (EU) 2019/947"

This Annex to the ED Decision contains 19 Articles pertaining to Acceptable Means of Compliance and Guidance Materials for unmanned aircraft operations outlined in CIR (EU) 2019/947. The sections are color-coded, with Guidance Materials in green and Acceptable Means of Compliance in yellow.

The Guidance Materials define and delineate:

- The terms "assemblies of people," "dangerous good," "autonomous operation," "uninvolved persons," "maximum take-off mass;"
- The boundaries between the categories of UAS operations;
- The requirements for UAS operations in the "certified" category;
- The minimum age for remote pilots;
- The rules for conducting an Operational Risk Assessment (specific to the JARUS SORA V2.0 process;
- Authorization of operations in the "specific" category;
- Cross-border operations or operations outside of the state of registration;
- UAS operations in the framework of model aircraft clubs and associations;
- Designation and tasks of the competent authority;
- Guidelines for risk-based oversight (by competent authorities); and
- The exchange of safety information and occurrence reports.

The list of Acceptable Means of Compliance addresses:

- The rules for conducting an Operational Risk Assessment;
- Predefined risk assessment ("PDRA");
- Authorizing operations in the "specific" category;
- Registration of UAS operators and "certified" UAS (including display of the registration; and
- Tasks of the competent authority.

This Annex does not apply to indoor UAS operations, which is defined as: "Indoor operations are operations that occur in or into a house or a building (dictionary defined) or, more generally, in or into a closed space such as a fuel tank, a silo, a cave or a mine where the likelihood of a UA escaping into the outside airspace is very low."

As stated elsewhere in this book, AMCs are a means of compliance, but not the only means, and GMs are for guidance, but are not regulatory requirements. However, material divergence from the terms and conditions may jeopardize the ability of a UAS operator to obtain permission to conduct operations in the European Union.

Opinion No 01/2020 High-Level Regulatory Framework for U-space

The audience for this lengthy Opinion includes Member States, unmanned aircraft systems operators, the manned aviation community, U-space service providers, air-navigation services providers, and all airspace users.

The stated objectives are to create and harmonize the necessary conditions for both manned and unmanned aircraft to operate safely in the U-space airspace, to prevent midair collisions, and to mitigate air and ground risks.

An enforceable and effective regulatory framework is necessary to support and enable operational, technical, and business developments, and to ensure fair access to all airspace users. Such a framework will stimulate the market to deliver U-space services that will respond to airspace users' needs.

This Opinion is the first regulatory step to allow immediate implementation of the U-space after the new Regulation (in an accelerated rule-making procedure, Rule-making Task RMT.0230) becomes effective so that UAS and U-space technologies can evolve. The Opinion contains a draft regulation that has been submitted to the European Commission. A related EASA decision (discussed below) contains acceptable means of compliance and guidance material. The Opinion proposes what EASA considers to be the minimum necessary rules, to be complemented in the future with additional provisions, enabling a more mature state of airspace integration. It is emphasized that the high-level regulatory framework is intended to allow immediate implementation of the U-space after the entry into force of the relevant Regulation.

The Opinion acknowledges that the full integration of the airspace used by manned and unmanned aircraft should be the goal for the future as the best solution to accommodate ATM as a whole, but the current state of the technology and the maturity (or lack thereof) of U-space services and detect and avoid systems do not allow for that level of integration. *"In summary, this is a first regulatory phase that is due to support operations as soon as the regulation is adopted and in the near future. It is focused on the principles of strategic and pre-tactical traffic management techniques (strategic because of the use of airspace management techniques to manage the U-space airspace, and pre-tactical because it is based on sharing information prior to and during flight."* (Emphasis added.)

The proposed rule structure is divided into eight chapters, or categories. The chapters are further divided into 25 Articles:

- Chapter I – Principles and general requirements
  - Article 1 Subject matter and scope
  - Article 2 Objectives
  - Article 3 Definitions
- Chapter II – Establishment of the U-space
  - Article 4 Designation of U-space airspace
  - Article 5 Common information service
- Chapter III – General requirements for aircraft operators and U-space service providers

- o Article 6 UAS operators
- o Article 7 Obligation for operators of manned aircraft operating in U-space air-space
- o Article 8 U-space service providers
- o Article 9 Occurrence reporting
- Chapter IV – U-space services
  - o Article 10 Network identification service
  - o Article 11 Geo-awareness service
  - o Article 12 Flight authorization service
  - o Article 13 Traffic information service
  - o Article 14 Tracking service
  - o Article 15 Weather information service
  - o Article 16 Conformance monitoring service
- Chapter V – CIS providers and U-space service providers
  - o Article 17 Application for a CIS provider and U-space service provider certificate
  - o Article 18 Conditions for obtaining a certificate
  - o Article 19 Validity of the certificate
- Chapter VI – Competent authorities
  - o Article 20 Competent authority
  - o Article 21 Tasks of the competent authorities
  - o Article 22 Exchange of safety information and safety measures
- Chapter VII – Pricing of CIS
  - o Article 23 Pricing of common information service
- Chapter VIII – Final provisions
  - o Article 24 Amendments to Commission Implementing Regulation (EU) 2017/373
  - o Article 25 Entry into force and applicability

The remainder of the Opinion presents a comprehensive narrative of the impact assessment EASA performed in consultation with various stakeholders, an analysis of the expected benefits and drawbacks of the proposed rules, all taking into account the European regulatory environment, EU policies, current UAS operations, international context, safety risk assessment, who is affected, how the issues and problems are expected to evolve, and what EASA wants to achieve with the new Regulation.

ED Decision 2020/022/R Amendment 1 to Acceptable Means of Compliance and Guidance Material to CIR 2019/947 and to the Annex (Part-UAS) Thereto, Issue 1, Amendment 1

This Executive Director Decision was issued on 15 December 2020. Its full title is "Amendment 1 to Acceptable Means of Compliance and Guidance Material to Commission Implementing Regulation (EU) 2019/947 and to the Annex (Part-UAS) thereto 'AMC and GM to Commission Implementing Regulation (EU) 2019/947 – Issue 1, Amendment 1' 'AMC and GM to Part-UAS – Issue 1, Amendment 1.'" This directive addresses the requirements set forth in Regulation (EU) 2018/1139 (the basic regulation that, among other things, established a European Union Aviation Safety Agency [EASA]), to issue certification specifications and acceptable means of compliance (AMC), as well as guidance material (GM), for the application of that Regulation (itself a comprehensive 122-page document).

This decision outlines the process whereby EASA developed the non-binding AMCs and GMs to demonstrate compliance with Regulation (EU) 2018/1139. In ED Decision 2019/21/R the Executive Director issued Acceptable Means of Compliance and Guidance Material to CIR (EI) 2019/947 and to Part-UAS thereof. Therefore, EASA shall reflect the state of the art and best practices in the field of aviation and update its decisions taking into account the worldwide aviation experience and scientific progress in the respective fields.

EASA accepted the JARUS proposal to evaluate the ground risk from operations over populated areas and assemblies of people, and created additional guidance on the application of risk mitigation means (through the JARUS/SORA process).

In addition, a format for registration of UAS operators and of certified UAS was developed that ensures interoperability of the national registration systems.

EASA has also developed new predefined risk assessments (PDRAs) to cover UAS operations that are proposed by Member States.

In light of the foregoing, Annex I and Annex II of ED Decision 2019/021/R are amended as set forth in Annexes I and II of this Decision (see below).

The Decision declares that AMCs and GMs are non-binding standards issued by EASA "which may be used by persons and organizations to demonstrate compliance with Regulation (EU) 2018/1139 and the delegated and implementing acts adopted on the basis thereof." Annexes I and II to Commission Implementing Regulation (EU) 2019/947 are amended to cover a broad range of topics pertaining to operational risk assessments, the JARUS SORA air conflict mitigation process, integrity and assurance levels for the operational safety objectives, predefined risk assessment (PDRA) methodology, and other matters mostly related to UAS and their operators, as opposed to air traffic management.

The primary concern of this Decision and its annexes (its specific objectives) are to:

1. Increase the number of types of operations covered under PDRAs, and facilitate harmonized operational authorizations across EASA Member States;
2. Facilitate interoperability of the national registration systems for UAS;
3. Promote mutual access to and exchange of information through a "broker solution" until the repository referred to in Article 74 of the Basic Regulation (Regulation 2018/1139) is established;
4. Increase safety, efficiency, and harmonization in the application of UAS regulations;
5. Foster the development of the EU UAS market;
6. Clarify the conditions under which UAS operations over populated areas and assemblies of people can be authorized in the "specific" category (which includes higher risk operations, where the drone operator obtaining an operational authorization from the national competent authority before starting the operation ensures safety by conducting a risk assessment); and
7. Achieve an acceptable level of safety and harmonization among EASA Member States, as well as facilitate societal acceptance of UAS operations in the "specific" category.

Each of these categories of mandates and goals has a direct impact on the EU's U-space integration plans. Full integration of UAS and manned aviation in non-segregated airspace is not achievable until every one of these goals has been realized.

Explanatory Note to Decision 2020/022/R

The objective of this Decision is to update the AMCs and GMs to Commission Implementing Regulation (EU) 2019/947 (the "UAS Regulation") and to the Annex (Part-UAS) thereto published with Decision 2019/021/R.

The amendments are expected to increase safety, improve harmonization among EASA Member States, and facilitate societal acceptance of UAS operations in the "specific" category.

The two subtasks clarify the conditions under which UAS operations over populated areas and assemblies of people can be authorized in the "specific" category, and ensure the interoperability of Member States' national registration systems for UAS operators and certified UAS that require registration. In addition, new PDRAs are introduced that are intended to improve existing PDRAs.

The objective of this Decision is to update the Acceptable Means of Compliance (AMC) and Guidance Material (GM) to Commission Implementing Regulation (EU) 2019/947 (the "UAS Regulation") and to the Annex (Part-UAS) thereto, as published with Decision 2019/021/R.

EASA developed this Decision under rulemaking task (RMT).0730, which is divided into the following two subtasks:

- Subtask 1a clarifies the conditions under which unmanned aircraft system (UAS) operations over populated areas and assemblies of people can be authorized in the "specific" category; and
- Subtask 1b ensures the interoperability of the national registration systems, which are established and maintained by the EASA Member States (MSs) for UAS operators and for certified UAS that require registration, introduce new predefined risk assessments (PDRAs), and improve the existing PDRAs.

The amendments are expected to increase safety, improve harmonization among EASA Member States, and facilitate societal acceptance of UAS operations in the "specific" category.

Annex I to ED Decision 2020/022/

This Annex refers to AMCs and GMs to Commission Implementing Regulation (EU) 2019/947, Issue 1, Amendment 1. Employing strikethroughs and color-coding to indicate new or amended text, the document references the JARUS SORA (V2.0) process to delineate rules for conducting an Operational Risk Assessment. As an AMC/GM document, the process described therein is not mandatory, however, as emphasized in other AMC/GM publications, if employed to assess risk of an operation or system, it must be strictly adhered to. The SORA process is a complex method of assessing operational risk, with many steps that present abundant opportunities for digressions that may lead to a denial of operational authorization from the competent authority with legal jurisdiction over the proposed operation or system.

This document is apparently intended to be user friendly, with colorful charts, graphs, and checklists to aid the proponent in preparing a credible risk assessment.

The topics covered in this Annex are:

- Rules for conducting an operational risk assessment;
- The SORA process (with graphs and flow charts to describe the steps to be taken);
- Strategic mitigation – collision risk assessment;
- Integrity and assurance levels for the operational safety objectives (OSOs);
- Pre-defined Risk Assessments (PDRAs);
- Personnel in charge of duties essential to the UAS operation;
- Registration of UAS operators and certified UAS.

Annex II to ED Decision 2020/022/R AMC and GM to the Annex (Part-uas) to Regulation (EU) 2019/021/R

This Annex contains templates for the application forms for operational authorizations and instructions for filling out those forms.

The scope of privileges accorded to LUC (Light UAS Certificate) holders are also listed. LUC holders are operators of small UAS in the "open" and "specific" categories.

The privileges may be one or more of the following:

- Conduct operations covered by standard scenarios without submitting the declaration;
- Self-authorize operations conducted by the drone operator and covered by a PDRA without applying for an authorization;
- Self-authorize all operations conducted by the drone operator without applying for an authorization.

No prior authorizations are required for commercial operations in the "open" category if the operations meet the "open" criteria. A UAS may be operated in the "specific" or "certified" categories when it does not meet the requirements laid out under the "open" category.

EASA Rules of the Air, Easy Access Rules for Standardized European Rules of the Air (SERA)

This 213-page document is also color-coded for ease of use (blue for implementing rules, orange for acceptable means of compliance, and green for guidance material). The document is updated regularly to incorporate amendments to the Regulations and Annexes.

The objective of this Regulation is to establish the common rules of the air and operational provisions regarding services and procedures in air navigation and are applicable to all airspace users and aircraft engaged in general air traffic operating into, within, or out of the European Union, as well as those bearing the nationality and registration marks of a Member State and operating in any airspace to the extent that they do not conflict with the rules of a country having jurisdiction over the territory overflown.

The Regulation also applies to competent authorities of the Member States, air navigation service providers (ANSPs) aerodrome operators, and ground personnel engaged in aircraft operations.

The Regulation specifically does not apply to model and toy aircraft, but Member States shall ensure that they establish national rules to ensure model and toy aircraft are operated in such a manner as to minimize hazards related to aviation safety, to persons, property, or other aircraft.

The Regulation defines 145 terms, including the meaning of "toy aircraft" (one intended for use in play by children under the age of 14), but does not define "unmanned aircraft" or "U-space." Neither term is mentioned anywhere in the Regulation, but other Regulations specific to unmanned systems and U-space refer back to SERA for the basic rules of the road in Europe.

Commission Implementing Regulation (EU) 2021/666 of 22 April 2021 Amending Regulation (EU) No. 923/2012 as Regards Requirements for Manned Aviation Operating in U-space Airspace

This regulation, which goes into effect on 26 January 2023, requires that manned aircraft operating in airspace designated by the competent authority as a U-space airspace, and not provided with air traffic control service by the ANSP, shall continuously make themselves electronically conspicuous to the U-space service provider. The regulation goes on to specify operational requirements for both VFR and IFR flights into radio mandatory and transponder mandatory zones.

Commission Implementing Regulation (EU) 2021/664 of 22 April 2021 on a Regulatory Framework for U-space

This Regulation consists of 19 Articles and 7 Annexes. The term "shall" appears throughout, which means that Member States and their competent authorities, U-space airspace, and common information service providers must comply with every requirement in the regulation, without exception, where the word "shall" appears. Some of the sections use "should" or "may" instead of "shall," rendering those elements as suggestions rather than requirements. U-space airspace service providers and UAS operators using the airspace must likewise comply with every relevant regulation.

The impetus for the regulation was the safety, security, privacy and environmental risks posed by the proliferation of UAS entering very low level (VLL) airspace, and the increasing complexity of beyond visual line of sight (BVLOS) operations. The European Commission, having previously established a first set of provisions for harmonized UAS operations and minimum technical requirements for UAS, determined there was a need to go further and define a specific set of rules for integrating UAS with manned aviation and ensuring safe separation between multiple UAS in defined geographical zones, which should be called "U-space."

The Regulation's articles and annexes impose minimum performance-based requirements for UAS operators and U-space service providers to facilitate the safety of operations in that airspace. The rules and procedures applicable to UAS operating in U-space should be proportionate to the nature and risk of the operations.

Very small UAS (maximum take-off mass of less than 250 g) operating within visual line of sight of the pilot are deemed to be low risk and not subject to these rules. Model aircraft (recreational or hobbyist) are likewise excluded from coverage due to their "good safety record." The regulations contemplate an operating partnership between UAS operators, U-space service providers, and air traffic service providers that are to operate under a harmonized set of rules for standardized services and connectivity methods.

Military and state aircraft are excluded from regulation, but Member States are encouraged to define static and dynamic U-space restrictions to allow for safe integration of those aircraft in U-space airspace.

Although not mandatory, the regulations acknowledge that Member States should be able to designate a single common information service provider that will provide common information services on an exclusive basis in some or all U-space airspaces under their jurisdiction.

At a minimum, U-space service providers should provide four mandatory U-space services, consisting of (1) network identification services, (2) geo-awareness services, (3) UAS flight authorization services, and (4) traffic information services. Supplementing those requirements is the rule that calls for effective signaling of the presence of manned aircraft through use of undefined surveillance technologies, as set forth in Commission Implementing Regulation (CIR) (EU) No. 923/2012, as amended by CIR (EU) 2021/666 (discussed below).

Other requirements of this groundbreaking regulation are described in the following paragraphs.

CHAPTER I

PRINCIPLES AND GENERAL REQUIREMENTS

Article 1: Subject Matter and Scope

The regulations apply to UAS operators, U-space service providers and common information service providers. They do *not* apply to operations conducted by model aircraft clubs and associations; subcategory A1 of the "open" category of operations as defined in other Implementing or Delegated regulations; and in accordance with SERA.5015 instrument flights rules of Implementing Regulation (EU) No. 923/2012.

Article 2: Definitions

This article sets forth the definitions of six terms unique to U-space operations in addition to those found in four other Implementing and Delegated Regulations.

CHAPTER II

U-SPACE AIRSPACE AND COMMON INFORMATION SERVICES

Article 3: U-space Airspace

Member States designating U-space airspace for safety, security, privacy, and environmental reasons shall support that decision with an airspace risk assessment (using the criteria set forth in Annex 1).

Where a Member State chooses to designate U-space airspace, all UAS operations in that airspace shall be subject to four mandatory U-space services (after performing an airspace risk assessment): (1) Network identifications services; (2) geo-awareness services; (3) UAS flight authorization services; and (4) traffic information services.

Member States have the option of requiring additional U-space services, based upon the airspace risk assessment. The risk assessment for all U-space airspace must determine UAS capabilities and performance requirements, U-space service performance requirements, and the applicable operational conditions and airspace constraints.

Member States shall give access to airspace service providers' relevant data from its UAS operators registration system, as well as registration systems from other Member States. U-space airspace information must be made available through the Member State's aeronautical information service and in accordance with Article 15(3) of EU Regulation 2019/947.

If a Member State decides to establish a cross-border U-space airspace, the Member States must jointly decide on the designation of the cross-border U-space airspace, and the provision of cross-border U-space airspace and common information services.

Article 4: Dynamic Airspace Reconfiguration

This Article deals with a Member State's designation of U-space airspace within controlled airspace. The dynamic reconfiguration of the airspace must conform to ATS.TR.237 of the Implementing Regulation 2021/665, amending Regulation (EU) 2017/373, to ensure that manned aircraft being provided with air traffic control services remain segregated from UAS.

Article 5: Common Information Services

The common information services that Member States must make available are: (1) Horizontal and vertical limits of the U-space airspace; (2) the requirements determined pursuant to Article 3(4); (3) a list of certified U-space service providers, along with their identification and contact details, the U-space services provided, and any certification limitations; any adjacent U-space; (4) UAS geographical zones relevant to U-space and published by the Member State in accordance with Implementing Regulation 2019/947; and (5) static and dynamic airspace restrictions defined by the relevant authorities that permanently or temporarily limit the volume of airspace within the U-space where UAS operations can occur.

Relevant operational and dynamic airspace reconfiguration data must be included as part of the common information services provided by the Member State's U-space airspace service providers.

U-space service providers must make the terms and conditions of their services available to the common information services in the airspace where they offer their services. The information provided must be made available consistent with Annex II and must comply with the data quality, latency, and security protection required in Annex III.

All common information services must be accessible to relevant authorities, air traffic providers, U-space service providers, and UAS operators on a non-discriminatory basis, ensuring the same data quality, latency, and security protection levels as stated above.

Member States may, but are not required, to designate a single common information service provider to supply the services described above on an exclusive basis, who must do so consistent with the requirements stated in previous paragraphs of this section. Such single common information service provider shall fulfill those same requirements and must be certified in accordance with Chapter V of this Regulation. The Member State that makes this designation must inform EASA and the other Member States of any decision concerning the certificate of the common information service provider. EASA will retain the information in the repository referred to in Article 47 of Regulation (EU) 2018/1139.

CHAPTER III

GENERAL REQUIREMENTS FOR UAS OPERATORS AND U-SPACE SERVICE PROVIDERS

Article 6: UAS Operators

UAS operators intending to use U-space airspace are compelled to abide by the capabilities and performance requirements and established by the Member State with jurisdiction over the desired U-space airspace, as set forth in Article 3(4)(a), and applicable operational conditions and airspace constraints as per Article 3(4)(c). Operators must utilize the mandatory U-space services listed in Article 3(2) (network identification, geo-awareness, flight authorization, and traffic information services) and Article 3(3), which provides for additional services that may be required by the Member State pursuant to Articles 12 (weather information service) and 13 (conformance monitoring) of this regulation.

UAS operators may provide their own U-space services, in which case they become U-space service providers subject to all the relevant requirements in this Regulation.

UAS operators may not enter U-space airspace unless they hold an operational authorization or certificate issued by the competent authority of the Member State of registration (not the State where operations are conducted) of the UAS pursuant to the requirements of Implementing Regulation (EU) 2019/947. In addition, the operator must comply with any operational limitations set by a Member State in any UAS geographical zone.

UAS operators must submit a flight authorization request to the U-space service provider through the UAS flight authorization service (Chapter IV, Article 10) before each individual flight.

The UAS operator may not commence the intended flight until it requests and receives activation of flight authorization from the relevant U-space service provider. The operator shall comply with the terms of the flight authorization and any deviation thresholds (as provided in Article 10(2)(d)). The service provider may introduce changes to the authorization at any time during any phase of the flight, which must be communicated to the operator, who must comply with the changes. The UAS operator must request a new flight authorization if they cannot comply with flight deviation thresholds.

Last, UAS operators shall provide for contingency measures and procedures, which in turn must be made available to U-space service providers.

Article 7: U-space Service Providers

U-space services must be provided by "legal persons certified as U-space service providers" in accordance with Chapter V. The awkward term "legal persons" presumably means those individuals or entities that are qualified to serve in that role and have successfully completed the certification process set forth in Chapter V. The regulations do not state whether the provider may be an individual, or a corporation, company or some other legal entity, such as a partnership or joint venture.

U-space service providers must provide all UAS operators with the airspace services described in Articles 3(2) and (3) during all phases of operations in that airspace. They must also establish arrangements with air traffic service providers in their designated airspaces to facilitate coordination of activities and the exchange of relevant operational data and information as required by Annex V.

Air traffic data must be handled without discrimination, restriction, or interference "irrespective of their sender or receiver, content, application or service, or terminal equipment." This presumably means that no authorized user of the U-space airspace will be given priority over another, regardless of the identity of the operator or the source of the data. Both U-space service providers and air traffic service providers must use a common, secure, and open communications protocol per Annex II.

The U-space service providers could be one entity handling all service functions or several entities working in concert to provide the whole menu of services provided in this Regulation. As such, providers must exchange amongst themselves any information relevant to the safe provision of services, must adhere to a common secure open communication protocol, and use the latest available information. The information exchanged must meet the data quality, latency, and protection requirements set forth in Annex III, and the providers must ensure unimpeded access to and necessary protection of the information that is exchanged.

U-space service providers must report to the competent authority at the commencement of operations after receiving their certificate per Article 14, and must likewise report ceasing and restart of operations if applicable.

CHAPTER IV

U-SPACE SERVICES

Article 8: Network Identification Service

Network identification services must provide for continuous remote identification services throughout the duration of a flight, which includes dissemination to authorized users "in an aggregated [collective] manner." Authorized users are defined as: (1) The general public regarding information deemed public in accordance with applicable (European) Union and national rules; (2) other U-space service providers; (3) air traffic service providers; (4) any designated single information service provider; and, (5) the relevant competent authorities.

The network identification service shall allow authorized users to receive messages with the following content: (1) UAS operator registration number; unique serial number of the UAS (or if home-built, the serial number of the "add-on"; (2) the geographical position of the UAS, its altitude above mean sea level and height above the surface or take-off point; (3) the route course measured clockwise from north and the UAS ground speed; (4) the geographical position of the remote pilot (or the UAS take-off point); (5) the emergency status of the UAS; and (6) the time that the message was generated. The information shall be updated at a frequency that is determined by the competent authority.

Article 9: Geo-awareness Service

The U-space service provider must provide geo-awareness information to UAS operators, consisting of operational conditions and airspace constraints within the U-space airspace, UAS geographical zones relevant to the U-space airspace, and any temporary restrictions applicable to the U-sairspace. This information must be communicated in a timely manner to allow for contingencies and emergencies to be addressed by the UAS operator, and must include the time of update and the version number or a valid time, or both.

Article 10: UAS Flight Authorization Service

U-space service providers are required to provide authorization services to UAS operators for each individual flight, setting the terms and conditions of that flight (similar to information exchanged between air traffic service providers and aircraft operators for IFR clearances). Upon receiving a flight authorization request, the service provider must: (1) ensure that the request is complete in accordance with Annex IV; (2) accept the request if the flight is free of intersection in space and time from any other notified flight authorization within the same airspace, in accordance with the priority rule in Section 8 (the eight special operations exceptions referenced in Article 4 of Implementing Regulation No. 923/2012 [Common Rules of the Air], primarily covering police, customs, firefighting, emergencies, and the like); (3) notify the UAS operator of the acceptance or rejection of their request; and (4) when notifying the UAS operator of acceptance of the authorization, indicate the flight authorization deviation thresholds.

In considering and issuing flight authorizations, the service provider must also utilize applicable weather information provided by the weather information service called, referred to in Article 12.

If the U-space service provider is unable to grant the authorization request, it may (but is not required to) offer an alternative flight authorization to the UAS operator.

The U-space service provider must confirm the activation of the UAS flight authorization without unjustified delay after receiving the request for authorization.

U-space service providers must make arrangements to resolve conflicting flight authorization requests received from UAS operators via different U-space service providers. (This requirement, number 6 in Article 10, is somewhat ambiguous, but the preamble to the Regulation suggests that the intention is for U-space service providers operating in the same geographical zone to establish a process whereby conflicts can be resolved, relying upon common information services and coordination among providers. There is no direction informing the providers how they are to go about making those arrangements, whether unilaterally or in coordination with the other providers).

U-space service providers shall "check" the request for UAS flight authorization against the airspace restrictions and temporary airspace limitations (presumably before granting an authorization). Implied, but not stated, is that authorization will not be granted if such restrictions or limitations are such as to present a hazard to navigation. In any case, U-space service providers must give priority to UAS conducting special operations referred to in Article 4 of Implementing Regulation (EU) No. 923/2012 (discussed above). If two authorization requests have the same priority, they shall be processed on a first-come-first-served basis.

Service providers must continually check existing flight authorizations against new dynamic airspace restrictions and limitations, as well as information about manned aircraft traffic shared by air traffic service units, with emphasis on manned aircraft known or believed to in a state of emergency (including unlawful interference, such as hijacking or forced diversion) and update or withdraw authorization as required by the circumstances.

U-space service providers shall issue a unique authorization number for each UAS flight authorization. The number shall enable identification of the flight and the identity of the UAS operator and the U-space service provider that issued the authorization.

Article 11: Traffic Information Service

The information service offered to UAS operators must contain information on any other "conspicuous" air traffic that may be in proximity to the position or intended route of the UAS flight. Information about manned aircraft and UAS traffic shared by other U-space service providers and air traffic service units must also be provided to UAS operators. Information provided about known traffic shall include position, time of report, speed, heading or direction, and emergency status of the aircraft, if known, and it must be updated at a frequency determined by the competent authority.

Upon receiving this information from the service provider, the UAS operator shall take the relevant action to avoid a collision hazard.

Article 12: Weather Information Service

U-space service providers shall collect weather data from reliable sources to maintain safety and support other U-space services, provide UAS operators with weather forecasts and current weather information, and the information shall include, at a minimum: (1) wind direction measured clockwise through true north; (2) speed, in knots per second; (3) height of the lowest broken or overcast layer in hundreds of feet above ground level; (4) visibility in meters and kilometers; (5) temperature and dew point; (6) indicators of convective activity and precipitation; (7) location and time of observation, or valid times and locations of the forecast; and (8) appropriate QNH (barometric pressure adjusted to sea level) with the geographical location of its applicability. The provided information must be up-to-date and reliable to support UAS operations (basically, the contents of a standard aviation weather briefing consistent with Annex 3 of the Convention on International Civil Aviation).

Article 13: Conformance Monitoring Service

This U-space service shall enable UAS operators to verify whether they are in compliance with the requirements set forth in Article 6(1) and the terms of the flight authorization. The service also has the mandate to inform a UAS operator when flight deviation thresholds are violated and when the Article 6(1) requirements are not being met. When a violation or deviation is detected, the service shall so inform other UAS operations in the same vicinity, other U-space service providers for the same airspace, and the relevant air traffic services providers.

CHAPTER V

CERTIFICATION OF U-SPACE SERVICE PROVIDERS AND SINGLE COMMON INFORMATION SERVICE PROVIDERS

Article 14: Application for a Certificate

This regulation describes the certification application process for a U-space service provider and, if designated, a common information service provider. The provider must hold a valid certificate issued by the competent authority of a Member State where it maintains its principal place of business.

If the provider's principal place or business or residence is in a third country (presumably not a European Union Member State), they must apply for certification with EASA (the "Agency").

The U-space service provider's certification shall be issued in accordance with Annex VI and the single common information service provider's certificate is issued in accordance with Annex VII. The certificate shall set forth the rights and privileges of its holder to provide services to which it relates.

The application for these two types of certifications, or for an amendment to an existing certificate, shall be on a form and manner established by the competent authority or by the Agency, whichever is applicable.

Article 15: Conditions for Obtaining a Certificate

This Article lists 12 categories of requirements for a U-space service provider and a single common information service provider to be certified to provide those services:

1. They must be able to provide services in a safe, secure, efficient, continuous, and sustainable manner consistent with the level of performance established by Member States pursuant to Article 3(4).
2. They must use systems and equipment that guarantees quality, latency, and protection of the U-space or common information services.
3. They must have appropriate net operating capital commensurate with the costs and risks of providing the sought-after services.
4. They must have the ability to report occurrences in accordance with ATM/ANS. OR.A.065 in Subpart A of Annex III to Implementing Regulation (EU) 2017/373 (Occurrence reporting).
5. Implement and maintain a management system in accordance with Subpart B of Annex III to Implementing Regulation (EU) 2017/373.
6. Implement and maintain a security management system in accordance with Subpart D of Annex III to Implementing Regulation (EU) 2017/373.
7. Retain recorded operational information and data for at least 30 days or longer, where the recordings may be pertinent to accidents and accident investigations, or until it is evident that the information is no longer required.
8. Create a robust business plan that shows that the provider can meet its obligations to provide continuous service for at least 12 months from the start of operations.
9. Have in place arrangements to cover liabilities related to the tasks, appropriate to the potential loss and damage (essentially a comprehensive liability insurance requirement).

10. Where the services of another service provider are required, agreements specifying the allocation of liability between them must be in place.
11. Develop a contingency plan to respond to events such as security breaches that may impact the delivery of services that result in significant degradation or interruption of operations.
12. In addition to the foregoing, service providers shall have an emergency management plan to assist a UAS operator experiencing an emergency and a communication plan to inform those concerned.

Article 16: Validity of the Certificate

A U-space service provider or single common information service provider's certificate remains valid as long as the holder of the certificate is in compliance with all the relevant requirements of this Regulation.

The certificate shall cease to be valid if the holder of the certificate has not started operations within 6 months after the certificate was issued or has ceased operations for more than 12 consecutive months.

The competent authority or the Agency shall continue to assess the operational and financial performance of any certificate holders for which it has responsibility. Based upon that assessment, the competent authority or the Agency may impose particular conditions to the certificate holder, amend, suspend, limit, or revoke the certificate.

CHAPTER VI

GENERAL AND FINAL PROVISIONS

Article 17: Capabilities of the Competent Authorities

Competent authorities shall have the technical and operational capacity to fulfill their respective obligations under Article 18 (below). They must have appropriately documented procedures and adequate resources, employ competent personnel with professional integrity, experience, and training to carry out their duties, and take any action required to contribute to the safe, efficient, and secure operation of UAS in U-space airspace under their responsibility.

Competent authorities shall be capable of initiating any appropriate enforcement measures necessary to ensure that the U-space service and common information service providers comply with the requirements of this Regulation.

Article 18: Tasks of the Competent Authorities

This Article specifies eleven tasks. They are:

1. Registration system: Establish, maintain, and make available a registration system for certified U-space service and information service providers.
2. Traffic data: Determine what traffic data, live or recorded, that U-space service, single common information service, and air traffic service providers must make available to authorized "natural or legal persons," including required frequency and quality level of data, without prejudice to personal data protection regulations.

3. Level of access: Determine the level of access to the information for users of the common information, in accordance with Annex II.
4. Data exchange: Ensure that data exchanged between air traffic and U-space service providers are made in accordance with Annex V.
5. Certificate application: Define the process for natural and legal persons to apply for U-space service or common information service provider certification.
6. Coordination mechanism: Establish a mechanism to coordinate with other authorities and entities, including those at the local level, U-space airspace designation, airspace restrictions within the U-space airspace, and determination of U-space services to be provided in the U-space airspace.
7. Certification and risk-based oversight programs: Establish a continuous certification and risk-based oversight program, which shall include monitoring of operational and financial performance, commensurate with the risk associated with the services to be provided by U-space service and common information service providers under the competent authority's oversight responsibility.
8. Information availability: Require that common information and U-space service providers make available all information necessary to ensure that provision of U-space services contribute to the safe operation of aircraft.
9. Oversight program: Conduct audits, assessments, investigations, and inspections of U-space service and common information service providers.
10. Risk assessment: Take into account required levels of safety performance when defining the requirements for each U-space airspace subject to an airspace risk assessment in accordance with Article 3(1).
11. Safety performance: Regularly monitor and assess the levels of safety performance and use the results as appropriate within their risk-based oversight responsibilities.

This Regulation shall apply from 26 January 2023.

Commission Implementing Regulation (EU) 2021/665 of 22 April 2021 Amending Implementing Regulation (EU) 2017/373 as Regards Requirements for Providers of Air Traffic Management/Air Navigation Services and Other Air Traffic Management Network Functions in the U-space Airspace Designated in Controlled Airspace

As suggested by the title, this Regulation amends language in Implementing Regulation (EU) 2017/373 by adding definitions of four terms unique to U-space operations and creating common requirements for coordination and notification protocols for air traffic service providers handling UAS traffic in U-space airspace contained within controlled airspace.

The new definitions are:

1. "U-space airspace" means a UAS geographical zone designated by Member States where UAS operations are only allowed to take place with the support of U-space services,
2. "U-space service" means a service relying on digital services and automation functions designed to support safe, secure, and efficient access to U-space for a large number of UAS,

3. "Common information service" means a service consisting of dissemination of static and dynamic data to enable the provision of U-space services for the management of unmanned aircraft traffic, and

4. "Dynamic airspace reconfiguration" means the temporary modification of U-space airspace in order to accommodate short-term changes in manned traffic demand, by adjusting the geographical limits of that U-space airspace.

Section 1 of Subpart A of Annex IV (of Implementing Regulation (EU) 2017/373) is amended at point "ATS.OR.127 Coordination by air traffic service providers in U-space airspace," to require air traffic service providers to: (1) Provide, on a non-discriminatory basis, relevant traffic information regarding manned aircraft that is necessary for a U-space airspace established in a controlled airspace where the air traffic service provider is the designated provider of U-space airspace service (per Implementing Regulation [EU] 2021/664, discussed above); and (2) establish the coordination procedures and communication facilities between appropriate air traffic service units, U-space service providers, and, where applicable, single common information service providers, permitting provision of this data.

Additional language is inserted in Section 2, Subpart B of Annex IV, point "ATS.TR.237 Dynamic reconfiguration of the U-space airspace," to read as follows:

"Air traffic control unit shall:

(a) Temporarily limit the area within the designated U-space airspace where UAS operations can take place in order to accommodate short term changes in manned traffic demand by adjusting the lateral and vertical limits of the U-space airspace, and

(b) Ensure that the relevant U-space service providers and, where applicable, single common information service providers are notified in a timely and effective manner of the activation, declaration and temporary limitations of the designated U-space airspace."

In addition to these three new Implementing Regulations, published simultaneously on 22 April 2021, and which do not go into effect until January of 2023, EASA released other materials intended to clarify its approach to airspace integration, and in particular the establishment of designated U-spaces in the European Union.

Opinion No. 01/2020 High-Level Regulatory Framework for the U-Space

This opinion presaged the release of Implementing Regulation (EU) 2021/664. Its Executive Summary stated:

> "The objective of this Opinion is to create and harmonize the necessary conditions for manned and unmanned aircraft to operate safely in the U-space airspace, to prevent collisions between aircraft and to mitigate the air and ground risks. Therefore, the U-space regulatory framework, supported by clear and simple rules, should permit safe aircraft operations in all areas and for all types of unmanned operations.
>
> This Opinion proposes an effective and enforceable regulatory framework to support and enable operational, technical and business developments, and provide fair

access to all airspace users, so that the market can drive the delivery of the U-space services to cater to the airspace users' needs.

This Opinion, is, therefore, a first regulatory step to allow immediate implementation of the U-space after the entry into force of the Regulation and to let the unmanned aircraft systems and U-space technologies evolve."

The Opinion is structured to inform the public that it contained a draft regulation and was to be submitted to the European Commission, which was to use it as a technical basis in order to prepare an EU regulation. A draft text for acceptable means of compliance and guidance material was also prepared, in anticipation that it would be published by EASA when the European Commission adopts the regulation described above (the high-level regulatory framework, released on 16 December 2021).

The Opinion covered no fewer than 27 topics listed under four main headings: "About this opinion;" "In summary – why and what;" "Impact assessment;" and "Proposed actions to support implementation."

"About this opinion" described the process of how the Opinion was developed and proposed the next steps.

"In summary – why and what" discussed why a regulatory framework for U-space is needed, the objectives of creating a U-space, a notional rule structure, an overview of the proposals, the stakeholders' views and outcomes of the consultations, and the expected benefits and drawbacks of the proposals.

By far the longest section was "Impact assessment." Topics discussed included the general regulatory and policy concerns, current UAS operations, and the international context. Issues such as safety risk assessment, regulatory harmonization, growth of the number of UAS operations, whether the existing air traffic management system can handle the increased demand, human factors and the interplay between humans and highly automated UAS operations, and the unique characteristics of unmanned systems are discussed at length. Two options for achieving the stated goals are no policy change at the EU level, leaving U-space implementation to be achieved by individual Member States, or development of a harmonized framework for the establishment of U-space across Europe. As documented in the three Implementing Regulations discussed above, the regulators chose the second option.

The impacts section outlined the methodology employed to assess the impacts, the assumptions underlying the assessment, the safety, economic, environmental, security, and social impacts, and, no less important, the impact on general aviation.

"Proposed actions to support implementation" included focused communication in AB (advisory bodies) meetings, establishment of a UAS implementation network, provision of clarifications in electronic communication tools between EASA and NAAs, detailed explanations on the EASA website, dedicated thematic workshops and sessions at EASA, a series of thematic events organized on the regional principle, or combinations of any of the foregoing.

Appendix 1 to the Opinion lists 16 questions that should be included in a UAS flight authorization request form. Appendix 2 provides a form for a Certificate for the U-space Service Provider. Appendix 3 offers a form for a Common Information Service (CIS) provider certificate that may be issued by a competent authority.

CIR (EU) 2020/1166 Amending Implementing Regulation 2019/947 as Regards Postponing the Date of Application for Standard Scenarios for Operations Executed in or Beyond the Visual Line of Sight

This Regulation was issued on 15 July 2021. It addresses the timeline set forth in CIR (EU) 2019/947 (discussed above) regarding the ability of Member States to accept declarations made by UAS operators in complying with one of the two standard scenarios (the conditions under which unmanned aircraft system (UAS) operations over populated areas and assemblies of people can be authorized in the "specific" category, and the interoperability of the national registration systems for UAS operators and for certified UAS that require registration, also introducing new predefined risk assessments). EASA expected that the harmonized standards addressing the requirements applicable to UAS classes C5 and C6 would be available by 2 December 2021. EASA acknowledges that those harmonized standards will not be available by that date, so the date has been extended to 3 December 2023. Therefore Article 5(5) of Implementing Regulation 2019/947 shall apply after that date and Member States can accept declarations made by UAS operators in accordance with Article 5(5), based on standard national scenarios or equivalents, until that date. Such declarations shall cease to be valid after 2 December 2025. Article 5(5) states: "Where the UAS operator submits a declaration to the competent authority of the Member State of registration in accordance with point UAS.SPEC.020 laid down in Part B of the Annex for an operation complying with a standard scenario as defined in Appendix 1 to that Annex, the UAS operator shall not be required to obtain an operational authorization in accordance with paragraphs 1 to 4 of this Article and the procedure laid down in paragraph 5 of Article 12 shall apply." Article 12 addresses authorizing operations in the "specific" category. That category covers operations presenting a higher risk that require a thorough risk assessment, as described elsewhere.

The rule here for UAS operators intending to conduct operations in the EU is that they will not have a designated U-space airspace in which to operate for at least another two to three years, so their only choice is to decide on the type of operation they want to conduct and try to abide by the complex matrix of regulations the European Union has implemented, as summarized in the preceding paragraphs.

In August of 2021 EASA released a **Notice of Proposed Amendment 2021-09 (RMT.0730)** as a regular update of acceptable means of compliance and guidance material to Regulation (EU) 2019/947 on the rules and procedures for the operation of unmanned aircraft. Pertinent to the topic of airspace integration and development of the U-space airspace, this amendment proposes to update and amend existing acceptable means of compliance and guidance materials for the rules and procedures for operating UAS in European airspace. Specifically, with reference to U-space, the NPA:

- Proposes new AMC and GM for the definition of "geographical zones." When an EASA Member State publishes maps to illustrate a geographical zone, it should ensure consistency with Chapter 8, "UAS geographical zone data model of EUROCAE ED-269." Common layouts and similar color codes should be used, especially for cross-border UAS operations, to maintain consistency among Member States. U-space airspace is an example of such a geographical zone. U-space is referenced no fewer than a dozen times throughout this NPA.

- Proposes a new AMC defining the procedures to be applied to UAS operators and the competent authorities to cross-border operations, which could include U-space airspace.
- Proposes multiple AMCs and GMs dealing with UAS operations in the "open" and "specific" categories, which would comprise the vast majority of operations in U-space airspace.

Ultimately, all of the proposed amendments and additions to existing AMCs and GMs pertaining to UAS operations in the European Union are works in progress, subject to regular review and amendment. The NPA acknowledges that the AMCs and GMs of the document only concern the first developmental phase in the creation of U-space airspace (static geographical zones in terms of time and location and predefined zones activated within a predefined time frame). The second two phases (dynamic geographical zones in terms of time that are activated/deactivated without preannouncement and dynamic geographical zones in terms of time and location) will be enabled through the forthcoming U-space regulatory framework (see Opinion No. 01/2020 High-level regulatory framework for the U-space, discussed above). Additional AMC and GM may need to be developed at that time.

Commission Implementing Regulation (EU) 2020/639 of 12 May 2020 Amending Implementing Regulation (EU) 2019/947 as Regards Standard Scenarios for Operations Executed in or Beyond the Visual Line of Sight

This CIR made material changes to 2019/947. It added 11 new definitions (unmanned aircraft observer; airspace observer; command unit (CU); C2 link service; flight geography; flight geography area; contingency volume; contingency area; operational volume; ground risk buffer; and "night").

Amendments or additions to Articles 5, 13, 14, 15, 22, and 23 are made, and a new Annex replaces the original Annex.

Paragraph 5 of Article 5 is amended by adding "The UAS operator shall use the declaration referred to in Appendix 2 to that Annex" at the end.

Most of the changes are reflective of new terminology and technology advances revealed by research and input from Member States, focused on Standard Scenarios 1 and 2.

Standard scenario 1 (STS-01) covers operations executed in visual line of sight (VLOS), at a maximum height of 120 m over a controlled ground area in a populated environment using a CE class C5 UAS.

Standard scenario 2 (STS-02) covers operations that could be conducted beyond visual line of sight (BVLOS), with the unmanned aircraft at a distance of not more than 2 km from the remote pilot with the presence of airspace observers, at a maximum height of 120 m over a controlled ground area in a sparsely populated environment, and using a CE class C6 UAS.

Also addressed are improvements to conspicuity, cross-border operations, and practical skill training and assessment of remote pilots.

Commission Implementing Regulation (EU) 2020/746 of 4 June 2020 Amending Implementing Regulation (EU) 2019/947 as Regards Postponing Dates of Application of Certain Measures in the Context of the COVID-19 Pandemic

This CIR does what the title suggests. The highlights are:

1. Measures introduced to contain the COVID-19 pandemic severely hamper the ability of Member States and the aviation industry to prepare for the application of a number of recently adopted Implementing Regulations in the field of aviation safety.
2. Confinement and changes in the working conditions and availability of employees combined with the additional workload required to manage the significant negative consequences of the COVID-19 pandemic for all stakeholders are impairing preparations for the application of these Implementing Regulations.
3. A delay in executing the different tasks required for the proper and timely implementation of Commission Implementing Regulation (EU) 2019/947, notably the establishment of registration systems that are digital and interoperable, as well as the adaptation of authorizations, declarations, and certifications issued on the basis of national law, is inevitable as a consequence of the COVID-19 pandemic.
4. The standardization process and other related activities led by the industry and standardization bodies, such as preparation of testing methodologies, or the testing of technical features, such as the remote identification, have been delayed. This in turn will have a negative impact on the capacity of manufacturers to put on the market unmanned aircraft systems (UAS) meeting the new standardized requirements in accordance with Commission Delegated Regulation (EU) 2019/945.
5. All UAS types should thus be allowed to continue to operate under the existing conditions for an additional 6 months. Therefore, the dates of application of Implementing Regulation (EU) 2019/947 should be postponed accordingly in order to allow UAS operators to be able to use UAS not complying with Delegated Regulation (EU) 2019/945 for an additional 6 months.
6. The European Union Aviation Safety Agency confirmed to the Commission that postponing the application of the provisions referred to in recital 3 is possible without having a detrimental effect on aviation safety, since it will be for a very limited period, the air traffic is likely to resume slowly in the context of recovery from the COVID-19 pandemic, resulting in less exposure to the risks, and national law will continue to apply in those Member States where UAS operations are permitted.

Commission Delegated Regulation (EU) 2020/1058 of 27 April 2020 Amending Delegated Regulation (EU) 2019/945 as Regards the Introduction of Two New Unmanned Aircraft Systems Classes

Again, the title reveals the focus of this regulation. The regulation imposes additional requirements for operations under the "open" and "specific" categories, also with reference to the "certified" category. To further advance the need for harmonization across jurisdictions, two new different classes of UAS should be created, characterized by different sets of requirements addressing different risks. Consequently, Chapter II of Commission Delegated Regulation (EU) 2019/94 should cover these new classes.

Where manufacturers place UAS on the market with the intention of making them available for operations under the rules and conditions applicable to the "open" category or under an operational declaration and therefore affix a class identification label on it, they should ensure the compliance of the UAS with the requirements of that class. Equally, where manufacturers place on the market accessories kits transforming a class C3 UAS into a C5, they should ensure compliance of the UAS fitted with the accessories kits with all the requirements of class C5.

In order to support the remote identification as one of the necessary elements for the functioning of the U-space system under development, all UAS operated in the "specific" category should be equipped with a remote identification system.

Those UAS operated in the specific category, and that are not required to register in accordance with Article 14 of Implementing Regulation (EU) 2019/947, should have a unique serial number, unless they are privately built.

This regulation outlines amendments to Articles 1–6, 8–9, 12–14, 16–17, 30, 36, and 40 of CIR 2019/945.

That covers EASA's regulatory response. We will now talk about the organizations that are laying the groundwork for implementation of U-space CONOPs.

Implementing Organizations

Sesar Ju

U-space; Supporting Safe and Secure Drone Operations in Europe; A Report of the Consolidated SESAR U-space Research and Innovation Results

In 2015, the European Commission mandated the creation of a Single European Sky Air Traffic Management Research Joint Undertaking (SESAR JU), whose role is to develop the new generation European air traffic management system and produce a blueprint on how to make drone use in low-level airspace safe, secure, and environmentally friendly. Subsequently, the SESAR JU received a further mandate to conduct research and demonstration activities on a wide variety of technical challenges, with the goal of transforming the resulting U-space Blueprint into reality.

Recognizing the huge potential of the growing drone ecosystem, in 2016 the European Commission launched U-space – an initiative aimed at ensuring the safe and secure management and integration of drones into the airspace. This set in motion a series of activities across Europe directed towards the development of appropriate rules and regulations, as well as technical and operational requirements, capable of supporting future autonomous operations.

In 2017, the European Commission chartered the SESAR JU to coordinate all research and development activities related to U-space and drone integration. In that same year the SESAR JU published its U-space Blueprint, setting out the vision and steps for the progressive deployment of U-space services from foundation services to fully integrated operations

(U1–U4). This was followed by the 2020 edition of the European ATM Master Plan, which incorporated a drone roadmap.

With the involvement of the European Union Aviation Safety Agency (EASA) and the European aviation industry standards developing body, EUROCAE, the projects also seek to ensure that their results can evolve within ongoing drone standardization and regulatory efforts.

The SESAR JU began with the publication of the U-space Blueprint, outlining the vision and notional steps for the progressive deployment of U-space services, from foundation services, such as registration, e-identification and geo-awareness, to more complex operations in dense airspace requiring greater levels of automation and connectivity. Building on the Blueprint, the SESAR JU then went further into detail with a roadmap for the safe integration of drones into all classes of airspace. This embeds not just the timeline for U-space, but it also outlines the steps to be taken to ensure a coordinated implementation of solutions to enable remotely piloted aircraft systems to fly alongside commercial aircraft. This roadmap was included in the 2020 edition of the European ATM Master Plan, which is the main planning tool shared by all stakeholders for air traffic management (ATM) modernization in Europe.

The projects were conducted in close coordination with the European Union Aviation Safety Agency, tasked by the Commission with drafting rules to govern the safe integration of drones into manned airspace and to help identify the operational requirements needed for this regulatory framework.

The initial effort focused on three service categories: U1 (Foundation Services); U2 (Initial services), and U3 (Advanced services). A fourth category, U4 (UAM/eVTOL, an extension of U3), was not included in the SESAR-JU research efforts. U3/U4 research activity anticipates the introduction of eVTOL (electric-powered vertical takeoff and landing aircraft, operated from "vertiports" in an Urban Air Mobility CONOPs). UAM refers to an integrated system that enables on-demand, highly automated, passenger or cargo-carrying air transport services, with particular reference to urban and suburban environments, where aviation activities are comprehensively regulated in most jurisdictions.

In 2017 and 2018, the SESAR JU launched 19 exploratory research projects and demonstration projects aimed at researching the range of services and technological capabilities needed to make U-space a reality. The projects brought together 25 European airports, 25 air navigation service providers, 11 universities, more than 65 start-ups and businesses, as well as 800 experts, working in close cooperation with standardization and regulatory bodies, including EUROCAE and EASA.

After two years of research, SESAR JU partners completed 19 research and demonstration projects. To ensure close cooperation with aviation standards developing organizations, the SESAR JU project team included international organizations such as EUROCAE (European Organization for Civil Aviation Equipment) and ICAO (particularly ICAO's Standards and Recommended Practices – SARPS). Recognizing the need to have a broader view on U-space, the projects also involved organizations representing new entrants, such as the Global UTM Association (GUTMA) and Drone Alliance Europe, as well as non-aviation entities from the telecommunications industry.

The project identified and researched a number of technical and operational categories deemed essential to the integration of unmanned systems into the national and international airspaces under the jurisdiction of the European Union. The CORUS CONOPS project (discussed below), having received broad consensus among its partners, developed a U-space architecture and provided detailed definitions of airspace categories to be used for VLL (very low level) drone operations, as well as the services that would be available within them to ensure safe and efficient operations.

The Consolidated Report (U-space Research Innovation Results 2020) details coverage and the maturity level of each of the designated U-space services (based upon the CORUS CONOPS third edition of 2019). This report purports to map the "state of play on development of the technological capabilities and services required for making U-space a reality, starting with foundation services (U1) before progressing to initial services (U2) and advanced services (U3)." As stated above, U4 was not addressed by these research activities.

U1 (Foundation Services): e-identification coverage; geo-awareness coverage; registration assistance coverage; registration coverage; and DAIM (drone aeronautical information management) coverage. All but DAIM are described at 100%, whereas DAIM coverage was 50%. Some services were only partially analyzed, primarily those identified with emergency management, and the research was limited to uncontrolled VLL airspace, with only one USSP working at a time.

U2 (Initial Services): the main block of services for U-space with respect to the services initially defined at the start of the research activities by the SESAR JU. These are the numbers indicating the maturity levels in each category:

- Legal recording (42%)
- Procedural interface with ATC (42%)
- Weather information (34%)
- Navigation coverage information (33%)
- Geospatial information (10%)
- Communication information coverage (4%)
- Electromagnetic interference information (0%)
- Population density map (0%)
- Geofence provision (100%)
- Monitoring (100%)
- Surveillance data exchange (100%)
- Tracking (100%)
- Traffic information (100%)
- Emergency management (6%)
- Navigation infrastructure monitoring (4%)
- Communication infrastructure monitoring (4%)
- Digital logbook (100%)
- Operation plan preparation/optimization (100%)
- Operation plan processing (100%)
- Incident/accident reporting (100%)
- Risk analysis assistance (100%)
- Strategic conflict resolution (100%)
- Reporting (0%)

U3 (Advanced Services) revealed: dynamic capacity management (100%); collaborative interface with ATC (7%); and tactical conflict resolution (3%). The report noted that the dynamic capacity management service was covered by several demonstration projects, whereas the other two categories considered a limited number of configurations. The "validation of U-space tests in urban and rural areas" (VUTURA) and the "safe and flexible integration of initial U-space services in a real environment" (SAFIR) projects addressed the tactical conflict resolution service, with limited, but useful, deliverables.

This report concludes that these research projects demonstrated that U1 and U2 services are currently ready for environments of low-level complexity (rural areas, segregated airspace, low traffic density), and that U2, U3, and U4 services (in more complex environments, such as cities and populated areas) must be further studied. Innovations and advances in technology, as well as new regulatory responses, will be necessary to support U3/U4 operations.

The major product of the research is the U-space Concept of Operations (CONOPs), providing an initial U-space architecture and description of airspace types and U-space services to enable safe and efficient very low-level (VLL) drone operations. U-space is described as an architecture of harmonized services and procedures relying on a high level of digitalization and automation of functions to support safe, efficient, equitable, and secure access to airspace for large numbers of drones. It provides an enabling framework to support routine drone operations and addresses all anticipated types of missions, including operations in and around airports. Ultimately, U-space is envisioned to facilitate complex drone operations with a high degree of automation to take place in a wide variety of operational scenarios.

CORUS (See also the discussion under EUROCONTROL)

The SESAR JU research and demonstration projects precipitated a series of workshops that drew hundreds of stakeholders, leading to the development of a Concept of Operation for EuRopean Unmanned Air Traffic Management Systems (CORUS), which was published in 2019 (referenced above).

Seeking a broader engagement in the aviation community, the European Commission launched the European Network of U-space Demonstrators in 2018, a forum to share knowledge and support the work of research bodies, such as the SESAR JU and regulatory agencies such as EASA. The resulting network extends the community and engages more stakeholders in the task of developing a robust framework for unmanned and manned aircraft to share the airspace.

The CORUS CONOPs, published in 2019, proposed a comprehensive picture of U-space airspace that can be clearly understood and may form a foundation on which implementation of U-space throughout Europe can be based. The CORUS-XUAM (Concept of Operations for European UTM Systems – Extension for Urban Air Mobility) project, formed out of CORUS, is focused on highlighting the importance of U-space solutions for integrated operations that require mutual situational awareness of traffic. The project is intended to demonstrate integrated operations of both manned and unmanned aircraft employing advanced U-space services and to explore ways to facilitate a proper interface with ATM/ANS (air navigation services), with a particular emphasis on airport operations.

The CORUS-XUAM project began with an update and expansion of the CORUS project's original U-space Concept of Operations. This project addresses the integration of UAM and drone operations into a more complex airspace and identifies new or emerging U3/U4 services. The project, launched on 1 December 1 2020, and, coordinated by the EUROCONTROL Experimental Centre in France, involves a 24-month, very large-scale demonstration (VLD), which will explain how U-space services and solutions could support integrated Urban Air Mobility (UAM) flight operations. "These services should allow electric vertical take-off and landing vehicles (eVTOL), unmanned aircraft systems (UAS) and other airspace users (unmanned and manned) to operate safely, securely, sustainably and efficiently in a controlled and fully integrated airspace, without undue impact on operations currently managed by ATM. The ultimate goal of the project is to develop safe, secure, environmentally responsible and acceptable UAM solutions deployed by 2025–2030 that will support seamless freight, emergency, security and mobility services."

The European ATM Master Plan Edition 2020 (referenced above) is the main planning tool for ATM modernization across Europe. The plan supports the Single European Skies (SES) effort and the larger EU aviation strategy and defines the development and deployment priorities needed to deliver the Single European Sky ATM Research (SESAR) vision. The Master Plan is regularly updated by SESAR, with collaboration between all ATM stakeholders, in order to respond to the evolving aviation landscape.

One major concern addressed in the plan is the explosive growth in the drone industry, which is creating new markets and innumerable business opportunities. This phenomenon is presenting new challenges to the aviation community, particularly in the realm of urban air mobility and service delivery. It also poses a significant and increasingly complex challenge to the ATM infrastructure, including regulatory oversight of the system.

In view of the anticipated large number and varied types, sizes, masses, and capabilities of these aircraft, the ATM system will require new forms of traffic management and air–ground system integration. Highly automated or autonomous aircraft (single-pilot operations, urban air mobility aircraft, cargo delivery drones, autonomous electrically powered air taxis, etc.) operators will want and demand equal access to the airspace. Simultaneously, renewed interest in the potential for operating remotely piloted or autonomous aircraft at very high altitudes (above 50,000 ft. MSL) will require access to and from the stratosphere through managed airspace. In Europe the perceived need for change is becoming even more urgent, as the limits of the current system, resulting in increasing delays and disruptions, can already be observed (notwithstanding the effects of the COVID-19 pandemic, which reduced air travel for over 18 months and is still ongoing as of this writing). At least one expert, the AeroDynamic Advisory Group, predicted in 2020 that the airline industry would not return to normalcy until late 2023. Others are not so optimistic. Ironically, the hiatus imposed by the pandemic may actually provide a much-needed respite from the pressure to optimize paths to full airspace integration, which is being driven by an increasing need or desire to enable new forms of flight operations that are attracting an impressive share of global investments.

In order to manage future traffic growth safely while mitigating the environmental impact, SESAR's vision is to deliver a fully scalable traffic management system capable of handling growing air traffic, both manned and unmanned. The vision builds on the notion of trajectory-based operations, which enable airspace users to fly their preferred flight trajectories, delivering passengers and goods on time to their destinations as cost-efficiently as possible.

The realization of this vision also depends on the integration of the wide variety of new unmanned aerial systems accessing the same airspace as conventional manned aircraft. As discussed at length above, this is the U-space, a framework designed to fast-track the development and deployment of a fully automated drone management system, in particular for, but not limited to, very low-level airspace. Scalable by design, U-space relies on high levels of autonomy and connectivity in combination with emerging technologies. Alongside the creation of U-space is the need to integrate large, remotely piloted aircraft systems into manned traffic, with special provisions designed to compensate for the lack of a pilot on board the aircraft to "see and avoid" or "detect and avoid" other aircraft or obstacles. The roadmap covering drone integration into U-space is incorporated into this 2020 edition of the Master Plan.

The Plan outlines a four-phase approach to solving the problems presented by the dynamics detailed in the SESAR JU and CORUS-XUAM documents, and others. Phase A is to "address known critical network performance deficiencies by delivering solutions that enhance collaboration between stakeholders, including across state borders and with aircraft, implementing initial system-wide information management, and introducing network capacity and demand balancing measures." Phase B envisions "efficient services and infrastructure delivery through the launch of first ATM data services, the introduction of cross-border free-route operations, and the integration of advanced airport performance management into the network and the provision of initial U-space services."

Phase C contemplates "defragmentation" of European skies through "virtualization and dynamic airspace configuration, supported by the gradual introduction of higher levels of automation support, the full integration of airports into ATM at network level and the management of routine drone operations."

Phase D envisions a "digital European sky through the delivery of a fully scalable system for manned and unmanned aviation supported by a digital ecosystem, full air-ground system integration, distributed data services, and high levels of automation and connectivity." The plan calls for delivery of a fully scalable system able to handle both manned and unmanned aviation, consistent with the joint industry declaration of "towards a digital sky," by 2040.

Linked to this, an ATM automation model that mirrors the five-level model from the Society of Automotive Engineers (SAE) (ranging from Level 0, "no automation," to Level 5, "full automation") was developed. The model illustrates the level of automation anticipated for each phase of the Master Plan (included in the last section discussing functional decomposition and referencing the Plan at page 13).

As described in the Plan, the current architecture of the European airspace system is restricted by limited capacity, poor scalability, fixed routes, and fixed national airspace structures at the airspace level, limited automation and low levels of information sharing at the air traffic service level, and a fragmented ATM infrastructure at the physical (sensors, infrastructure) level.

The notional future architecture proposed by the Plan includes, at the top level, high-altitude airspace operations, network operations requiring dynamic and cross-FIR (flight information region) airspace configuration and management, free routes and greater flexibility and resilience in responding to disruptions, and changing demand. Air traffic services would be upgraded and enhanced by automation support and visualization and scalable capacity. Data and application services and U-space operations will depend upon unified information and U-space interfaces. In addition, at the foundation of the system, is an integrated and rationalized ATM infrastructure.

The roadmap for the safe integration of drones into all classes of airspace was adopted by the SESAR Administrative Board in March 2018. The document outlines which drone-related research and development (R&D) activities should be prioritized in order to support the expansion of the drone market and achieve the smooth, safe, and fair integration of these new aircraft systems into the European airspace. A core element of that goal is the development and implementation of U-space, the framework designed to fast-track the development and implementation of a drone management system. The immediate target is VLL airspace operations, but the focus is not limited to that domain, as provisions to integrate large remotely piloted aircraft systems into manned traffic are also under consideration. The key elements of this roadmap were incorporated into this latest edition of the Master Plan.

The Plan proposes three foundation services: electronic registration, electronic identification, and geo-fencing and recommends a phased rollout of these services. Supportive services would include UAS geo-zone management with UTM (including data exchange interfaces with manned aviation stakeholders and ATM service providers).

EPAS (European Plan for Aviation Safety 2021–2025 – EPAS)

The rule-making system in Europe on its face is cumbersome, but probably necessary, given the equal voice afforded to each individual Member State (27 in all) in the development of rules and standards. Nevertheless, EASA has been cautious in proposing and implementing UAS regulations beyond the basic regulation discussed above, and has not addressed U-space or UTM/ATM beyond the framework for a rule outlined in Opinion No. 01/2020 (High-Level Regulatory Framework for the U-space).

The European Plan of Aviation Safety (EPAS) is advertised as "the regional aviation safety plan for EASA Member States, setting out the strategic priorities, strategic enablers and main risks affecting the European aviation system and the necessary actions to mitigate those risks and to further improve aviation safety. EPAS is a five-year plan that is constantly being reviewed and improved, and updated on a yearly basis. The plan is an integral part of EASA's work program and is developed by EASA in close consultation with the EASA Member States and industry."

"The main objective of EPAS is to further improve aviation safety and environmental protection throughout Europe, while ensuring a level playing field, as well as efficiency and proportionality in regulatory processes. EPAS' aspirational safety goal is to achieve constant safety improvement within a growing aviation industry."

EASA is a non-voting member of the FAA's ARAC (Aviation Rulemaking Advisory Group), and it is invited and decides on a case-by-case basis whether to join specific task groups of these committees.

The impact the COVID-19 pandemic has had on aviation regulations in general and U-space and UTM airspace integration in particular has yet to be fully realized. The slow development of UTM and U-space regulations, standards, and procedures by States around the world has impeded the implementation of UTM/U-space programs since early 2020, and continues to the present day.

Nations that were the most aggressive in development of UTM/U-space regulations and policies retreated from operational UTM deployment programs beginning in about 2018 due to a number of reasons unrelated to the pandemic. Uncertainty over regulatory developments and standards and a lack of maturity in key technologies essential to autonomous BVLOS flights are the more obvious reasons. The COVID-19 pandemic has also slowed the capacity of regulators and air navigation service providers (ANSPs) to develop new programs and investments in areas that rely on income from traditional commercial aviation sources.

As indicated in the foregoing discussions, during the first half of 2021 the Federal Aviation Administration in the US and the European Commission began implementing the next series of high-level UTM/U-space regulations. "There are still areas of uncertainty in terms of technology maturity, standards and regulations required before fleets of autonomous drones flying beyond-visual-line of sight missions above people can be authorized – but at least we have a framework for investment."

EASA's European Plan for Aviation Safety 2021–2025 Volume I

Volume I begins by describing efforts EASA has made since spring of 2020, to manage unanticipated responses to the COVID-19 pandemic in returning aviation to normal operations after the unprecedented collapse of activity in the second quarter of the year. All aspects of aviation have been affected by the pandemic, and the many programs devoted to integrating unmanned aircraft systems into national airspaces are no exception.

Section 3.1.3.2 of Volume I provides a concise summary of the EU's regulatory efforts towards UAS integration and establishment of a U-space (detailed above). The highlights of the EU regulatory response were the establishment of three categories of UAS operations ("open," "specific," and "certified"); recognizing the need for an unmanned traffic management system (U-space); the release of updated acceptable means of compliance and guidance materials that were published with Opinion No. 01/2018; the AMC and GM included specific operations risk assessment (SORA) as AMC to the risk assessment that is required in the "specific" category; and the first predefined risk assessment to assist operators when applying for an authorization in the "specific" category.

Key actions taken included:

- A decision amending the AMC and GM to define the risk classification for UAS operations in the "specific" category, operating in an urban environment, and to introduce additional predefined risk assessments.
- A first NPA for UAS in the "certified" category addressing all aviation domains (initial airworthiness, continuing airworthiness, remote pilot licenses, aircraft operations, rules of the air, ATM/ANS, and aerodromes) as well as VTOL operations: the NPA was expected in Q2/2021 and is planned for UAS operations in an urban environment (under RMT.0230).
- Opinion No. 01/2020, on a high-level regulatory framework for the U-space, was published in March 2020 (under RMT.0230) and the EC was to submit a revised draft regulation at the EASA Committee in Q1/2021.

Other topics addressed in Volume I included an Action Plan for Counter UAS (C-UAS) led by an EASA Task Force comprised of Member States (including NAAs and Law Enforcement Authorities), aerodrome operators, aircraft operators, ANSPs, EUROCONTROL, and the European Commission. The C-UAS Action Plan listed five Objectives and Deliverables. As of the publication of the EPAS document, only one deliverable (publication of safety promotion material) has been completed.

In addition, two paragraphs were devoted to "new air mobility" (another way of describing urban air mobility, or the introduction of new business models such as eVTOL air taxis). Although not specifically directed to unmanned systems, other innovations that affect low-level airspace management and integration of new systems into non-segregated airspace are electric and hybrid propulsion and vertical take-off and landing aircraft that will share the airspace with drones of all sizes and configurations.

Rule-making for design requirements that will address airworthiness standards and environmental issues is required to identify regulatory gaps in the existing regulations with regard to electric and hybrid propulsion systems, which could also include remotely piloted or autonomous aircraft powered by those same systems. This task will be done through RMT.0731 (New Air Mobility) for continued airworthiness requirements for all aircraft; and RMT.0230 (Drones), addressing manned e-VTOL electric propulsion aspects related to the ADR (aerodrome), ATM (air traffic management), FCL (flight crew licensing), and OPS (operations) domains (RMT stands for Rulemaking Task).

This publication noted that the greatest impact of the pandemic on drone operations in Europe was extending the effective date of the drone Regulation (EU) 2019/947 to 31 December 2022, giving drone operators another six months of breathing room to comply with the new regulation. CIR (EU) 2020/1166, discussed above, extended the applicability date to 3 December 2023. EPAS Volume I goes on to acknowledge that, while the pandemic has not stopped drone activity, development of UTM systems and applications is still a work in progress, and drone operators continue to mature conceptual frameworks, platform architectures, methodologies, and practical demonstrations. Rather than an impediment to progress, the COVID-19 pandemic may actually have accelerated the use of drones in humanitarian applications such as delivery of vital supplies and aid to medical personnel.

The EPAS plan outlines Europe's vision for using the ATM Master Plan to improve and modernize air traffic management across Europe. The European Commission recently issued an amended proposal on the implementation of the Single European Sky (SES II + recast), which proposes upgrading the whole SES framework of the EU Basic Regulation (which includes treatment of the emerging drone community and the need for a U-space plan to accommodate the new technology). It can be assumed that all efforts to upgrade the regulatory system and bring the technology capability level of the system up to a higher level will include integration of unmanned aircraft into that system.

EASA's European Plan for Aviation Safety 2021–2025 Volume II

The structure of Volume II reflects the various domains defined within the SRM process to provide a link with the corresponding safety data portfolios included in the ASR. The structure also facilitates the identification of actions relevant for different stakeholder groups:

- All systemic safety and competence of personnel issues are grouped within Chapter 5, which is further subdivided into seven distinct sections to address the various action areas.
- All actions other than those related to systemic safety and competence of personnel, corresponding to drivers' "safety," "level playing field," and/or "efficiency/proportionality" are grouped per domain (see Chapters 6 to 15). Within each of those chapters, actions are grouped per driver. For the driver "safety," a further grouping per key risk area is applied where a significant number of actions is included (this concerns Chapters 6 and 8 mainly).
- Regular update RMTs are included in the respective domain chapter.
- All actions corresponding to the driver "environment" are included in a separate Chapter 16.

Chapter 14 of Volume II (Unmanned aircraft systems) comprises all the actions that EASA deems relevant to ensure the safe integration of civil unmanned aircraft systems into the aviation system. Most of the EU Member States have adopted their own national regulations to ensure safe operations of UASs with a maximum take-off weight of 150 kg. The extension of the EU's scope of authority through the Basic Regulation to regulate UAS with MTOWs below 150 kg, and the recent adoption of EU requirements for the operations of UAS in the "open" and "specific" categories (CIR (EU) 2019/247 and 2019/945) compels Member States to modify their already adopted national regulations. The ultimate goal of these actions is to enable harmonized regulations at the EU level. They are also linked with other actions in EPAS (such as RMT.0731) that are intended to enable standardized UAS operations, as well as more complex operations of UASs such as operations in an urban environment (e.g. urban air mobility).

"In order to ensure safe UAS operations, it is extremely important to manage the safe integration of UASs into the airspace. U-space is a set of new services and specific procedures designed to support the safe, efficient and secure access to airspace for large numbers of drones. In 2017, the SJU prepared the U-space Blue Print describing the vision for U-space. In addition, the European Roadmap for safe integration of drones in all airspace classes was also prepared by the SJU with EASA support and adopted by the EC. The ATM MP reflects the details about the integration of UASs into the EU airspace."

The actions to be taken to enable the safe integration of drones in the European airspace, while maintaining a high and uniform level of safety, are described below:

RMT.0230 Introduction of a Regulatory Framework for the Operation of Drones

This Rulemaking Task includes development of Implementing Rules for UASs, implementing Articles 55 to 57 of and Annex IX to the Basic Regulation (IR (EU) 2018/1139). Articles 55 to 57 pertain to unmanned aircraft.

This task will also cover the development of AMC & GM to support the U-space regulation. Three categories of UAS are defined:

- The "Open" category defines a low-risk operation not requiring authorization or declaration before flight.
- "The Specific" category defines medium-risk operations requiring authorization or declaration before flight.
- The "Certified" category contemplates high-risk operations requiring a certification process.

Seven subtasks are identified. They include developing regulations of "open" and "specific" categories by implementing and delegating acts; creating specific regulations for certain operational scenarios in the "certified" category; introducing standard scenarios for "open" and "specific" categories (covered by RMT.0729); development of AMC & GM to support U-space regulation; and further amendments to other domains relating to introduction of detect and avoid systems and capabilities.

RMT.0729 Regular Update of Regulations (EU) 2019/945 & 2019/947 (Drones in the "Open" and "Specific" Categories)

The task includes the addition of standard scenarios (STSs) in Appendix 1 to the Annex to Regulation (EU) 2019/947, defining the conditions when a UAS operator can start an operation after having submitted a declaration to the competent authority. Also, the inclusion of new Parts in the Annex to Regulation (EU) 2019/945, including the technical requirements that UAS need to meet in order to be operated in the STSs, and establishing two new UAS classes (C5 and C6). Subtasks will cover two standard scenarios: VLOS in urban environments over controlled areas and BVLOS in sparsely populated environments over controlled areas using visual observers.

The task will also explore general improvements of Regulations (EU) 2019/947 and (EU) 2019/945. The second subtask will be activated when the need for amendment is identified.

RMT.0730 Regular Update of the AMC & GM to Regulations (EU) 2019/945 & 2019/947 (Drones in the "Open" and "Specific" Categories)

This task will develop predefined risk assessments (PDRAs) and recognition of industry standards in support of the specific operations risk assessment (SORA) methodology.

General improvements of AMC & GM to Regulations (EU) 2019/947 and (EU) 2019/945 will also be developed.

Subtask 1 updates the SORA process to accommodate BVLOS operations in urban environments, the development of three PDRAs, two copying the two standard scenarios developed in RMT.0729 (discussed above) and the other covering BVLOS operations over sparsely populated areas at less than 150 m above the surface and in uncontrolled airspace.

Subtasks 2 and 3 are focused on development of additional PDRAs, new AMCs & GMs for the definition of geographical zones, general improvement of AMC & GM, and recognition of industry standards.

Four additional ongoing efforts relevant to UAS airspace integration are also listed in the document:

1. **RES.0022 SESAR 2020 research projects aiming to safely integrate drones in the airspace** include surface operations by UAS (PJ.03a-09) and IFR UAS Integration (PJ. 10-05).
2. **RES.0023 SESAR exploratory projects on U-space** were launched by SESAR JU as a step towards realizing the European Commission's U-space vision for ensuring safe and secure access to airspace for drones.

3. **SPT.0091 European safety promotion on civil drones** seeks to promote safe operations of drones to the general public in furtherance of the overall goal of integration of RPAS/drones into European airspace.

4. **RES.0015 Vulnerability of manned aircraft to drone strikes** is an assessment of the collision threats to manned aircraft posed by drones. This will include evaluation of the estimated impacts of a drone and establishment of a risk model to support regulatory and operational responses that can be validated by a comprehensive set of simulated impact tests. (See also a parallel effort in the US conducted by the ASSURE organization and supported by an ASTM-developed standard, discussed in Chapter 6.)

A fifth rulemaking task related to the topic of airspace management and integration of new technologies into civil airspace, although not specifically mentioning drones or U-space, is **RMT.0731 New air mobility**. This task acknowledges that the introduction of new technologies and air transport concepts, such as multimodal and autonomous vehicles, requires revisiting the existing regulatory framework for aviation safety that was designed for conventional fixed wing aircraft, rotorcraft, balloons, and sailplanes. That framework relies on active involvement of human beings, often assisted by automation on board or on the ground. Continuing airworthiness requirements for electric and hybrid propulsion are under review and revision.

eVTOL electrically propelled, autonomously piloted aircraft are already being tested in the US and elsewhere. In September 2021 Airbus unveiled the latest generation of its CityAirbus electric air taxi, NextGen. This remotely piloted aircraft is equipped with fixed wings, a V-shaped tail, and eight electrically powered ducted propellers as part of its distributed propulsion system. It is designed to carry up to four passengers in a zero-emissions flight in multiple applications. These aircraft will add a new layer of complexity to the airspace integration/new air mobility problem, namely by being remotely piloted, and by carrying passengers.

EASA's European Plan for Aviation Safety 2021–2025 Volume III

Volume III of EPAS aims to present how aviation safety risks in Europe are analyzed and the outcome of these analyses (identifying where the risks are), with the goal of providing readers with more insight into the actions taken or recommended by EPAS.

This inaugural edition of EPAS Volume III provides the first insights to EASA's Safety Risk Portfolios. In their most simplified versions, the Safety Risk Portfolios are a list of safety issues that need to be mitigated at the European level.

These Portfolios have been developed by EASA in conjunction with its safety partners, through the CAGs (Collaborative Analysis Groups) and the NoA (Network of Analysts).

Safety Risk Portfolios form an essential component of EASA's SRM (Safety Risk Management) process. They gather insights from quantitative sources (e.g. data from the European Central Repository) and qualitative sources (e.g. expert judgment from safety partners). The combination of data analysis and expert judgment guides the strategic focus on the mitigation of safety issues that have been judged to be of high priority. Safety issues are identified through EASA's analysis of aviation occurrence data or submitted as a candidate safety issue through the CAGs, NoA, EASA's website, or internal EASA stakeholders.

The safety issues and Safety Risk Portfolios are grouped by domain as each domain has its specificities and requires specific expertise. The following domains are part of the SRM process:

- Aerodromes and Ground Handling
- ATM/ANS
- Commercial Air Transport – Aeroplanes
- Human Factors
- Non-Commercial Operations – Small Aeroplanes and rotorcraft

The 10 key risk areas identified by EASA's research are:

1. Airborne collision
2. Aircraft upset
3. Collision on runway
4. Excursion
5. Fire, smoke, and pressurization
6. Ground damage
7. Obstacle collision in flight (obstacles rising from the surface of the Earth)
8. Other injuries (an occurrence where fatal or non-fatal injuries have been inflicted that cannot be attributed to another Key Risk area)
9. Security (an act of unlawful interference against civil aviation)
10. Terrain collision

Arguably, the risk areas that would potentially involve integration of UAS into civil airspace or U-space would be airborne collision, aircraft upset, excursion, obstacle collision in flight, other injuries, security, and terrain collision. There are approximately 150 specific categories of risk listed in Volume III, and not all have direct relevance to unmanned aircraft operations or airspace integration. The particular Safety Issues that may have a direct impact on integration of UAS into national airspace or within designated U-space are discussed in the following sections.

Restarting a Complex System Is Challenging (SI-5005)

The aviation system is highly interconnected, sophisticated, and made up of people and technology, meaning that the consequences of shutdown and restart are not completely predictable. Organizations will need to prepare good communications and decision-making strategies, using personnel expertise, data, information, and good internal and external coordination.

The interconnectivity and coordination that will be required to safely manage a U-space could be compromised by a global degradation of air traffic demand and services caused by an event such as the COVID-19 pandemic or the outbreak of a regional armed conflict between nations that are part of the U-space system.

Cybersecurity (SI-2013)

ATM systems have become increasingly digitalized to reap efficiency gains. However, a move towards the digital sphere exposes ATM systems to more vulnerabilities and threats to confidentiality, integrity, and availability of the systems. Given the strong interdependence of the different domains in the aviation industry, a cyberattack on ATM systems may compromise safety and integrity of the aviation system as a whole. In addition to terrorist-related attacks, the safety issue is concerned with how ATM systems can remain resilient in the face of attacks perpetrated by hackers to gain access to systems or cause disruption for non-terrorist purposes and attacks carried out for commercial espionage.

As UAS are integrated into the notional U-space system, their network security mitigations, particularly those that intersect or connect to national or European air traffic management systems (EUROCONTROL) will need to be sufficiently robust to withstand a cyberattack, or be capable of implementing an impenetrable firewall between the U-space construct and the larger ATM systems.

Deconfliction between IFR and VFR Traffic (SI-4009)

Ineffective deconfliction of flights adhering to IFR and VFR in an airspace class where at least one of the flights is not under ATC separation has been identified as a strong contributor to airborne collision risk. Such airspace classes include class E, controlled airspace where VFR flights are not subject to ATC clearance and no IFR-VFR separation is provided by ATC, and class G, where neither IFR flights nor VFR flights are subject to ATC clearance and ATC does not provide any separation service. The safety issue arises due to the fragmented knowledge of the traffic situation as some traffic is subject to ATC clearance (i.e. IFR) and some traffic is not (i.e. VFR). ATC may not be aware of VFR flights or their intentions and potentially may not pass traffic information to the IFR traffic. In addition, some of the VFR traffic may not be equipped with ACAS or even a transponder (Mode-C or -S), reducing the conspicuity of VFR traffic. As a result, both IFR and VFR traffic have to rely solely on the visual acquisition by the flight crew to maintain separation. This safety issue addresses how the conspicuity of VFR traffic can be improved, as well as best practices to underscore the importance of existing procedures in maintaining airborne separation. This safety issue is captured in the Non-Commercial Operations – Small Aeroplanes Safety Risk Portfolio – and is also relevant to the Commercial Air Transport – Aeroplanes domain.

This issue becomes important when considering the EASA plan to integrate sUAS with manned aircraft operating in the U-space. As noted in the discussion of Easy Access Rules for Unmanned Aircraft Systems (Regulations (EU) 2019/947 and (EU) 2019/945, above), the European Commission issued three Implementing Regulations having a direct impact on airspace integration efforts in European skies, with specific focus on remotely piloted aircraft and U-space airspace. These regulations anticipate that RPAs will operate in some designated airspaces along with manned aircraft, and generally require that manned aircraft entering that designated airspace will be required to make themselves conspicuous to the U-space service providers and UAS operators so as to avoid conflicts and hazardous proximity to one another.

The introduction of a U-space, which calls for integration of manned and unmanned traffic in the same block of airspace, adds another layer of complexity to air traffic management strategies. The impetus for the Commission Implementing Regulation (EU) 2021/664 of 22 April 2021 on a regulatory framework for U-space was the safety, security, privacy, and environmental risks posed by the introduction of UAS into very low-level airspace, and the increasing complexity of BVLOS operations. The European Commission previously established a first set of requirements for harmonized UAS operations and minimum technical requirements for UAS. The Commission subsequently identified a need to go beyond the first tranche of UAS-related regulations and circumscribe a specific set of rules for integrating UAS with manned aviation and ensuring safe separation between multiple UAS in defined geographical zones, which is labeled "U-space."

These regulations anticipate an operating partnership between UAS operators, U-space service, and air traffic service providers that will operate under a harmonized set of rules for standardized services and connectivity methods. The focus of this research task is to ascertain how that partnership will work in a general sense, and should necessarily include U-space services as the effort evolves.

Effectiveness of Safety Management System (SI-2026)

Ineffective implementation of Safety Management Systems may lead to deficient management of ATM/ANS risks within the service provider organizations. The complex nature of aviation safety and the significance of addressing HF (human factors) aspects justify the need for effective management of safety by aviation organizations. Shared understanding between regulatory/competent authorities and air navigation service providers is imperative for an effective SMS functioning in an already ultra-safe industry like aviation. However, the lack of competent and experienced inspectors and strong regulatory authorities leads to the risk of bureaucratizing SMS, seeing it only as a compliance system. This safety issue covers the regulatory requirements and promotion of SMS principles, on both aviation authorities and organizations, and the capability to detect and anticipate new emerging threats and associated challenges.

The concept of "shared understanding between regulatory/competent authorities and air navigation service providers" is codified in "Commission Implementing Regulation (EU) 2021/664 of 22 April 2021 on a regulatory framework for U-space," discussed above. Those entities are bound by a very specific set of regulatory requirements for coordination and information sharing.

Failure of Air-ground Communication Service (SI-2018)

Failure of the air-ground communication system may degrade the performance of the communication service and increase safety risk to an unacceptable level. This safety issue explores how such failures can be prevented using pre-emptive measures and the best practices to manage such failures on a tactical basis when it occurs.

As U-space is created and cross-border operations or operations of manned and unmanned aircraft into, out of, or through designated U-space columns or corridors are authorized, management of air to ground communications systems will become increasingly complex, calling for innovative solutions. A key mode of an air–ground communication service is controller pilot data link communication (CPDLC), which allows air traffic controllers to transmit non-urgent strategic messages to an aircraft as an alternative to voice communications. Common failures in CPDLC include technical failure of data link equipment (air and ground) and disconnections known as "provider aborts."

Since there may be no air–ground voice communication between air traffic controllers and remotely piloted aircraft, automated systems may be required to fill the communications gap. This again invokes the human factors question of what communications that are traditionally handled by people may reasonably and safely be delegated to autonomous systems.

Failure of Navigation Service (SI-2016)

Failure of the navigation service may degrade the performance of the communication service and increase the safety risk to an unacceptable level. Air navigation service refers to the process of planning, recording, and controlling the movement of an aircraft from one place to another by providing accurate, reliable, and seamless position determination capability. Effective management of these services is essential in minimizing the impact on air traffic services (ATS). This safety issue covers appropriate maintenance procedures to identify failures and their impact on ATS, procedures to operate in degraded modes of operation, and training of staff to deal with abnormal situations.

Failure of Surveillance Service (SI-2017)

Failure of the surveillance service may degrade the performance of the communication service and increase safety risk to an unacceptable level. Surveillance systems are used by air traffic control to determine the position of aircraft. Such systems include secondary surveillance radar (SSR), the global navigation satellite system (GNSS), and Automatic Dependent Surveillance – Broadcast (ADS-B). Effective management of these systems is essential in minimizing the impact on air traffic services. This safety issue covers appropriate maintenance, procedures to identify failures and their impact on ATS, procedures to operate in degraded modes of operation, and training staff to deal with abnormal situations.

The U-space architecture contemplates a set of federated services and associated functions within a complete framework designed to enable and support safe and efficient multiple simultaneous drone operations in all classes of airspace. These services can be provided by different providers but such service providers will need to interoperate to performance requirements that are yet to be defined. The need to guarantee a seamless and safe operational environment will necessitate timely and accurate data transmission between implementation systems (Source: SESAR principles for U-space architecture).

With the declarations of Warsaw, Helsinki, and Amsterdam, the European Commission wants to create a competitive U-space services market for the benefit of the final users. This implies that the architecture allows multiple U-space service providers to operate in the same volume of airspace at the same moment. The architecture must then ensure that all the U-space service providers have exactly the same situational awareness and the traffic is de-conflicted (i.e. strategic or tactical deconfliction). This will require cooperation and exchange of data between the various service providers; connectivity and interoperability of the U-space services and related systems will then be essential.

However, the nature of some services is so safety or security and data privacy critical that they might be required to be unique and neutrally/centrally provided (e.g. registration, identification, geo-awareness, interface with ATM). The architecture has to allow this as well.

Finally, the U-space services will evolve to enable the growth in the number and variety of drone operations, supported by an appropriate interface with ATM. As time goes on, the whole aviation environment is expected to evolve into a fully integrated system supporting manned and unmanned operations in all classes of airspace.

Thus, to achieve the goals set forth by SESAR JU and the European Commission, operational integrity of a key component of that federated system must be assured. Surveillance services are a critical element of U-space architecture.

Integration of RPAS/drones (SI-2014)

This, of course, is the theme of this book. The exponential growth of drones, especially those weighing less than 25 kg and operating in the "open" category, has led to an increase of airborne collision risk between drones and manned aircraft. This is largely due to unauthorized activity of drones in both take-off and approach paths of commercial airlines up to 5000 ft. A potential result of a drone sighting is to stop or divert aerodrome traffic, leading to secondary risks, such as fuel exhaustion, airspace capacity saturation, and an increased workload for air traffic controllers and pilots. This research track falls under the heading of "ATM/ANS" in Volume III, almost as an afterthought, but it may well be the broadest in scope of all the EPAS research efforts. The authors of the document could just as well have designated this topic as a stand-alone research need, because the multi-layered aspects of the drone integration challenge compel an in-depth analysis that will have many subsets of inquiry.

New Technologies and Automation (SI-2015)

This safety issue refers to the potential increase in safety risks due to the complexities arising from the introduction of new technology and concepts in ATM, such as remote tower operations and system-wide information management (SWIM). With more complex automation, it is important to address the relationship between humans and automation within the framework of a contemporary safety management system. (More in-depth discussion of this topic will follow in Chapter 7).

Provision of Weather Information (Wind at Low Height) (SI-2009)

The landing phase is considered one of the highest-risk phases of flight due to the high cockpit workload and execution of difficult tasks such as the landing flare. Weather information near the surface of the runway, such as tail wind on the ground and cross wind, is crucial to assist flight crew during the landing phase. Inaccurate weather information may contribute to non-stabilized approaches and increase the risk of runway excursions. As this topic spans across several aviation domains, the scope of this safety issue is focused on the ANSP's and ATC's role in ensuring accurate and timely weather information is provided to flight crew during the landing phase.

U-space and UAM airspace domains envision very low-level operations, in both controlled and uncontrolled airspace. The need for timely and accurate weather information is one of the key components of the U-space architecture (see High-level regulatory framework for U-space, Chapter IV, Article 15, Weather Information Service, discussed above).

Effectiveness of Safety Management (SI-0041)

Aviation organizations are required to implement Safety Management Systems as part of their safety programs. The complex nature of aviation safety and the significance of addressing human factor aspects show the need for an effective management of safety by aviation organizations. The issue covers the regulatory requirements and promotion of SMS principles, for both aviation authorities and organizations, and the capability to detect, anticipate, and act upon new emerging threats and associated challenges. It also includes the settling of the adequate safety culture in organizations and authorities.

Plans for full integration of UAS into European airspace and U-space airspace create entrepreneurial opportunities for an entirely new class of participants, namely UAS operators without previous manned aviation experience, network identification services, geo-awareness services, UAS flight authorization services, traffic information services, weather information services, conformance monitoring services, and services providing effective signaling of the presence of manned aircraft through the use of still undefined surveillance technologies. It is critical to the overall level of safety and confidence in the new airspace system that every service or organization that plays an operational role in the system has a robust SMS program and safety culture so as not to introduce a single point of failure into the mix.

Decision-making in Complex Systems (SI-3016)

Decision-making in aviation activities can be complex, pressured, and bear a high risk. This, by definition, means that assessing trade-offs and interdependencies or making the right decisions can be difficult. Structures and processes to support decision-making can be helpful; however, the complexity of the system means that it is difficult to create such structures and processes with the necessary level of detail.

UAS integration and U-space management are also by definition complex problems. Among the many challenges that the regulators, developers, and operators must face is the issue of how much of the new system can or should be automated (or autonomous) and what core functions can only be trusted to humans. This is a decades-old focus of applied research in the aviation field, and has become even more significant in the rapidly evolving domain of unmanned aircraft systems. Much work by standards development organizations (SDOs) is being done in this realm (ASTM, IEEE, SAE, RTCA, ISO, etc., discussed in Chapter 6).

Design and Use of Procedures (SI-3007)

Procedures are used throughout the aviation industry to describe the correct actions and sequence of actions to perform a task. Out of necessity, procedures are designed using assumptions about the circumstances in which they will be applied. While this frequently produces well-designed procedures, the complex nature of the aviation working environment means that not every circumstance can reasonably be accounted for. Regardless of whether the procedure has been designed well or badly, rapid changes in the aviation system can mean that a procedure becomes more difficult to use over time.

This research topic clearly has application to UAS and airspace integration. Procedures are at the core of safe aviation operations (checklists, pre-flights, air traffic management processes, communications protocols, and so on), and all the more so for remotely piloted systems where the RPIC cannot "see" the environment in which the aircraft is operating, relying upon input from a variety of sources, all of which must be coordinated and harmonized in a manner that minimizes confusion and conflict. For fully autonomous operations (whatever that will come to mean), procedures must be built into the algorithms that control the system. Machine learning, human--machine interfaces, human-in-the-loop vs. human-on-the-loop, and other elements of robotic systems all depend upon certain assumptions and procedures that make the system work.

Impact of Culture on Human Performance (SI-3002)

Organizational culture is an important element in supporting human performance in the workplace. Culture depends on the historical context and the socio-technical environment and economic context in which we live. For example, with the "economic survival" effect, or when the "commercial benefit" dictates the running of the organization too much, this can lead to: a lack of resources; a stressful environment; no training policy; too much operational pressure and time pressure; too many subcontracting activities; insufficient maintenance or airport or ATC equipment; and so on (see the discussion of the Space Shuttle *Challenger* disaster in Chapter 7).

Organizational culture is no less as important in organizations devoted to unmanned aircraft and regulated airspace where those aircraft operate. Humans will still make decisions of strategy and tactics in the UAS ecosystem, driven by market forces, availability of technology, regulatory oversight and change, and financial strength or limitation. An organizational culture motivated more by short-term profit than long-term stability and quality of product or service could lead to the introduction into the larger system of an inadequate or unsafe product that could negatively impact the safety of the airspace and the public (see the discussion of the Boeing 737 Max debacle in Chapter 7).

Integration of Practical HF/HP into the Organization's Management System (SI-3004)

An organization is made up of humans, procedures, and processes, which work together, often in a hierarchical manner and interacting to achieve a common goal. As such, the organization's management system cannot be fully effective unless it has integrated HF considerations and human performance principles in a practical manner.

Along the same line of thinking in SI-3002, the outcome of this research will readily flow into safety cultures, organizational stability, and system safety in entities involved in UAS airspace integration research and development.

Knowledge Development and Sharing (SI-3008)

Knowledge sharing, particularly of tacit knowledge, is difficult to do well. This makes knowledge retention in situations of increased staff turnover very difficult. Knowledge development and sharing is about developing the right knowledge and making this knowledge available to the right people at the right time.

The sharing of information among participants in the U-space system is a regulatory requirement. Where a Member State chooses to designate U-space airspace, all UAS operations in that airspace shall be subject to four mandatory U-space services (after performing an airspace risk assessment): (1) Network identifications services; (2) geo-awareness services; (3) UAS flight authorization services; and (4) traffic information services (**Commission Implementing Regulation (EU) 2021/664 of 22 April 2021 on a regulatory framework for U-space**).

Article 5 (Common Information Services) of that regulation states that: The common information services that Member States must make available are: (1) Horizontal and vertical limits of the U-space airspace; (2) the requirements determined pursuant to Article 3(4); (3) a list of certified U-space service providers, along with their identification and contact details, the U-space services provided, and any certification limitations; any adjacent U-space; (4) UAS geographical zones relevant to U-space and published by the Member State in accordance with Implementing Regulation 2019/947; and (5) static and dynamic airspace restrictions defined by the relevant authorities that permanently or temporarily limit the volume of airspace within the U-space where UAS operations can occur.

Relevant operational and dynamic airspace reconfiguration data must be included as part of the common information services provided by the Member State's U-space airspace service providers.

U-space service providers must make the terms and conditions of their services available to the common information services in the airspace where they offer their services. The information provided must be made available consistent with Annex II and must comply with the data quality, latency, and security protection required in Annex III.

All common information services must be accessible to relevant authorities, air traffic providers, U-space service providers, and UAS operators on a non-discriminatory basis, ensuring the same data quality, latency, and security protection levels as stated above.

The results of this research task could have an impact on how the Member States should comply with the CIR (EU) 2021/664.

Organizational and Individual Resilience (SI-3009)

Organizational resilience is a key factor in successfully managing safe operation, but there is scant regulatory guidance on how to apply the concept. Resilience comprises both a system's ability to withstand disturbances, challenges and change, and to recover and sustain operations thereafter. The positive contribution to safety of each and every staff member is a key component in an organization's resilience.

The evolution of UAS airspace integration in general and design of the U-space in particular comprise many moving parts, or nodes, each independently managed but operationally interconnected. The ability of an organization or an individual in that organization to recover from a disruption of the system is one of the elements of a SORA-based risk assessment or Fault Hazard Analysis (also known as a Functional Hazard Analysis – more on that in Chapter 7). At the organizational level, each of the services that can or must be provided by a Member State participating in the U-space system must have strategies in place to adapt and respond to systemic changes such as the loss of authorization to provide that service. For individuals within those organizations who are engaged at the operational level, contingency management is an essential quality to maintain a safe environment. A computer glitch or perhaps an internet failure could bring down the whole system and create an unreasonable safety hazard (a type of single-point failure). Resilience and adaptability are core qualities for a safe aviation management system.

Aeroplane System Reliability (SI-4012)

The reliability and handling of any hardware/software system on board the airplane is crucial for a safe flight. This includes all systems on board the aircraft such as the aircraft structure, engine(s), flight controls, FMS, software incorporated into the system, etc. Failure of any of these hardware/software systems can result in loss of control and aircraft upset.

This safety issue is obviously important to remotely piloted aircraft, especially so because many of the smaller UAS that will operate in U-space will not have airworthiness certificates, so system integrity is critical to aviation safety if these aircraft are to occupy the same airspace as manned aircraft.

Damage Tolerance to UAS Collisions (SI-4019)

UASs are a growing airborne collision threat to manned aircraft due to their growing popularity among the public, who may not be aware of UAS regulations. It is important to consider the structural tolerance of a general aviation aircraft to withstand impact with a UAS and its ability to maintain controllability to enable a safe landing after a collision with a UAS. The damage tolerance has a direct relationship with the weight and size of the UAS.

This research could augment ongoing research into impact kinetic energy and what size, configuration, and mass limits a UAS may have to be integrated into civil airspace. Research efforts focused on the UAS rather than resilience of manned aircraft to midair collisions with drones, and has been authorized and conducted by various research organizations and universities around the world, in partnership with a number of civil aviation authorities.

We will return to some of these issues in Chapter 7.

A short note on "geographical zones," as defined in these regulations, is necessary to answer questions that may be posed by UAS operators in European very low-level airspaces.

All states are required to publish maps identifying geographical zones where all drone flights are forbidden or where you need to have a flight authorization before starting the operation. In most of Member States, apps for mobile phones are available to easily identify where UAS can operate.

Flight authorizations are different from the operational authorization required for the specific category to be flown. A flight authorization is applicable to all operations in an "open" or "specific" category and is issued by the authority/entity identified in the maps by the Member State. For example, a state may want to restrict the flights over a natural park or a riskier area, such as an industrial area or over a prison, etc. The state may then publish a geographical zone map requiring that all drone operations conducted in these zones must have a flight authorization issued by the authority managing the area (the park authority or the owner of the industry, for example).

Other types of geographical zones are those where one or more of the limitations of the open category are alleviated. For example, the state may authorize an area where all drones can operate up to a height of more than 120 m or with drones heavier than 25 kg or in BVLOS, etc., without the need for an authorization or a declaration. This may be very useful to model aircraft enthusiasts. Geographical zones should always be confirmed before starting the operation and the pilot/operator must always respect them (EASA Drones (UAS) *Regulatory reference Article 15 of EU Regulation 2019/947*).

Eurocontrol

EUROCONTROL's activities touch on operations, service provision, concept development, research, Europe-wide project implementation, performance improvements, coordination with key aviation players at various levels as well as providing support to the future evolution and strategic orientations of aviation.

EUROCONTROL defines U-space as "the European eco-system of services and functionalities to support drone operations. When fully deployed, a wide range of U-space missions that are currently being restricted will be possible. However, there is a need to develop a traffic management system for UAS, and define how it will all work technically and institutionally. The overall ATM system will need to handle low-level urban drone operations, high-flying military remotely piloted aircraft systems and the traditional mix of airlines, military, business and private jets. EUROCONTROL's role is to ensure the safe integration of UAS while safeguarding the rights of all airspace users."

The use of RPAS at lower altitudes is now a driving force for economic developments. Many of these smaller RPAS operate at altitudes below 500 ft AGL. According to ICAO Annex 2, this is the lowest available VFR altitude, and thus creates a possible boundary between smaller RPAS and manned aircraft. However, nearly every State allows manned operations below this altitude and coexisting with small undetectable RPAS poses a safety challenge. For now, no restrictions have been put in place regarding the maximum number of small RPAS allowed to operate in a certain area.

EUROCONTROL plays a pivotal role in developing European and global concepts, providing support to development of regulations for integrating manned and unmanned air vehicles, both civil and military, in the airspace and keeping a close eye on the wider UAS development landscape.

EUROCONTROL has developed a UAS ATM Integration Operational Concept, which is fully complementary to the EASA CONOPs (discussed above). EUROCONTROL develops safety cases and runs simulations that can assess the safety of drone operations, as well as reveal the complexities of integration. It possesses data on RPAS performance and has developed models capable of flying at different levels with different performances. Full implementation of this CONOPs is targeted after 2023, when the set of documents, rules, and technologies will enable seamless and safe integration of RPAS into ATM. The CONOPs does not describe or address different detailed scenarios, but provides an operational ATM perspective based on three areas of operation: very low level; between 500 ft AGL and FL600; and very high-level operations (above FL600). EUROCONTROL published its RPAS ATM CONOPs document, version 4, in 2017.

Performance requirements for RPAS operating in the same airspace as transport category aircraft include airspeed, latency, turn performance, and climb/descent performance. The overall approach towards ensuring safe RPAS integration is that RPAS has to fit into the

ATM system, not that the ATM system needs to be adapted to RPAS. Certain fundamental general integration requirements are:

- The integration of RPAS shall not imply a significant impact on the current users of the airspace;
- RPAS shall comply with existing and future regulations and procedures;
- RPAS integration shall not compromise existing aviation safety levels nor increase risk;
- The way RPAS operations are conducted shall be equivalent to that of manned aircraft, as much as possible;
- RPAS must be transparent (alike) to ATC and other airspace users.

Due to the absence of regulation and industry standards, accommodation of IFR-capable RPAS in controlled airspace is, for the time being, only possible through FUA/AFUA (flexible use of airspace/advanced flexible use of airspace) techniques. This is a daily occurrence in Europe for military RPAS. This phase of accommodation can easily be maintained due to the relatively low number of military RPAS operations. It is expected that the essential civil SARPS will be in place by 2023, which will enable civil and military RPAS to fly in non-segregated airspace.

After 2023 the anticipated availability of regulations, standards, and relevant supporting technology will permit RPAS, when meeting the specific airspace requirements based on the principles explained above, to integrate as any other airspace user.

The CONOPs breaks down the process of airspace integration to include a comprehensive airspace assessment. The list of topics includes:

- Increase of operations
- Introduction of BVLOS operations
- Safety concerns
- Environmental aspects
- Airspace classification
- Traffic complexity and density
- Zoning areas (hospitals, heliports)
- Geographic situation (mountains, urban areas)
- Traffic flows
- Noise
- Privacy
- Security
- Traffic forecast
- No drone zones
- Limited drone zones
- Segregated routes

The RPAS system description covers remotely piloted aircraft; remote pilot stations; C2 data link; ATC communications and surveillance equipment; ADS-B and secondary surveillance radar (SSR) transponder; navigation equipment; launch and recovery equipment; flight control computers (FCC); flight managements systems (FMS) and autopilot; system health monitoring equipment; and flight termination systems.

Types of operations addressed are: very high level (VHL) operations; IFR/VFR operations; very low level (VLL) operations (including both VLOS and BVLOS operations); and transition of manned operations below 500 ft.

This CONOPs proposes to organize RPAS traffic into classes. Each proposed class of RPAS traffic shall be implemented with all elements and requirements, as described. Implementation of individual elements will not be able to support safe integration RPAS into ATM.

The VLL system must consider RPAS flight planning; RPAS flight authorization; real-time RPAS tracking capability; provision of actual weather and aeronautical information; planned flight plans; active RPAS flight plans; airspace data; NOTAMS; weather; infra-structure availability; geo-fencing; and manned aircraft operations below 500 ft.

The CONOPs assumes that a C2 service is provided, the Member State has executed an airspace assessment, and that geofencing is in place or available, RPAS have surveillance capability similar in terms of performance and are compatible with manned aircraft surveillance capability, and a specific RPAS ATM-like management system is in place.

Four VLL traffic classes are proposed:

- Class I: Reserved for RPAS (EASA cat. A VLOS only). The buy and fly category that will be able to fly in low-risk environments and remains clear of no-drone zones like airports.
  - Mandatory declaration of operation
  - RPAS must be capable of self-separation in 3D
  - VLOS operations only
  - Geofencing capability, which ensures that this category remains separated from no-drone zones
- Class II: Free flight (VLOS and BVLOS). Can be the specific or certified category (EASA CONOPS).
  - Mandatory authorization for operation
  - Surveillance capability (4 G chip or other means)
  - VLOS and BVLOS operations
  - Free flight capability
  - RPAS must be capable of self-separation in 3D
  - BVLOS will have barometric measurement equipment
- Class III: Free flight or structured commercial route for medium/long haul traffic (BVLOS). Could be both specific and certified capable of operating for longer distances.
  - Mandatory authorization for operation
  - Surveillance capability
  - BVLOS operations only
  - Free flight or route structure
  - Shall have barometric measurement equipment
  - Can operate from ground up to 500 ft
- Class IV: Special operations (this category of RPAS traffic conducts very specific types of operation that will be assessed on a case-by-case basis (VLOS and BVLOS).
  - Require special authorization
  - Should be addressed on case-by-case basis
  - VLOS & BVLOS
  - Could require surveillance capability

Operational conceptual options are divided into three categories, depending upon the specificities identified in the airspace assessment. The first option contemplates the present situation not requiring an airspace assessment, such as around airports, nuclear power facilities, and other no-drone zones like hospitals, etc., falling into Classes I, II, and III. The second option is where RPAS traffic has increased to a point where a better-defined architecture is required. Traffic complexity and density can still allow free flight in Classes II and III, but Class I traffic will be restricted to lower altitude limits (150 ft), where higher traffic volumes are found. The detect and avoid capability becomes important to create a "bubble" around the RPA. Airspace assessment will be required to identify RPAS traffic flows to help define geographical areas where Class I operations will be restricted.

The third option imagines route structures as an alternative to the second option, with respect to supporting higher traffic demands. An airspace assessment can identify areas of low impact, such as railroad lines, rivers, and other geographical areas where there is minimal impact on people on the ground. DAA system requirements could be lower because of less complicated conflict encounter models than those invoked in the free flight category.

Moving on to the IFR/VFR Operations between 500 ft and FL600, the CONOPs offers three traffic categories that can operate in all traffic classes.

Class V traffic is IFR/VFR operations outside the "Network" (airport environment for IFR) and not flying SIDs (standard instrument departure route) or STARs (standard terminal arrival routes). RPAS not meeting Network performance requirements will be able to operate without negatively impacting manned aviation. Operations at airports will be accommodated through segregation of launch and recovery. Operations from uncontrolled airports or dedicated launch and recovery sites are to be conducted initially under VLOS/VFR until establishing radio contact with ATC. No additional performance requirements will be imposed in this environment, as compared to manned aviation.

General requirements for this class of operation are to file a flight plan that includes the following information:

- Type of RPAS
- Planned operations (navigation, route, flight level, etc.)
- Contingency procedure
- Contact phone number
- RPAS will meet CNS airspace requirements
- RPAS will be able to establish two-way communication with ATC if required
- RPAS will remain clear of manned aircraft
- RPAS operator must be able to contact ATC (if required) in regard to special conditions such as:
  o data link loss
  o emergency
  o controlled termination of flight
- RPAS D&A capability will be compatible and cooperative with existing ACAS system

Class VI is for IFR operations, including Network, TMA (terminal maneuvering/control area), and airport operations with RPAS capable of flying SIDs and STARs. These aircraft are capable of operating either as manned or unmanned platforms, meeting the requirements to operate in those areas. The general requirements are the same as for Class V, with the exception of the ability to remain well clear of other aircraft.

Operations of small RPAS above 500 ft are generally not allowed unless they meet IFR/VFR airspace requirements and can satisfy conspicuity requirements so as to be visible to manned aircraft. Member States can allow such operations on a case-by-case basis if the risk assessment of the intended operation is acceptably low.

Finally, very high-level operations (above FL600) are covered in Class VII, which consists only of IFR operations above FL600 and transiting non-segregated airspace. Although their operations will not directly impact the lower airspace, they will have to transit through either segregated or non-segregated airspace to enter or exit the airspace above FL 600. For such cases, temporary segregated airspace should be considered. Transition performance in segregated or non-segregated airspace below FL600 will be very limited since they will be focusing on long missions (up to several months). This airspace is mostly uncontrolled, so there ordinarily would be no traffic management requirement. However, the CONOPs mentions Google and Facebook as potentially operating around 18 000 aircraft at that flight level, from balloons to supersonic platforms, so some form of traffic management may become necessary.

General requirements for Class VII operations are:

- RPAS must file a flight plan.
- RPAS will meet CNS airspace requirements.
- RPAS must inform the responsible ATC unit in case of emergency re-entry into controlled airspace.
- RPAS must inform ATC about the type of contingency procedures to be used (balloon deflating or orbiting descent).
- A regional centralized system should have an overview of the ongoing operations.
- Departure and arrival procedures should be developed.

The flight plan should include the type of RPAS, contingency procedures, planned operation (navigation, route, flight level, etc.), and contact phone number.

In summary, this CONOPs is intended to be consistent with and conform to related efforts conducted by EASA, SESAR-JU, and ICAO. All of these organizations, and many others, collaborate and share concepts and ideas, with the primary goal of fully integrating UAS into all levels and classes of European airspace. These undertakings are extensive, iterative, and fluid as conditions change and technology advances. Regular updates to these published documents are an essential element of the global strategy towards harmonization and transparency. This technology sector is extraordinarily dynamic, and regular visits to the organizations' websites are necessary to stay up to date on regulatory and policy changes.

As we have seen through examining this CONOPs, which is now four years old and due for an update as a result of the looming introduction of eVTOL manned and remotely piloted "air taxis," EUROCONTROL's work focuses primarily on ATM-critical issues related to UAS integration that will maintain safety of ATM, deliver performance benefits, while monitoring new developments and ensuring that the non-ATM issues are properly identified and addressed by the relevant stakeholders in good time. Advanced Air Mobility and Urban Air Mobility concepts (see the discussion on SESAR-JU's CORUS and CORUS-

XUAM research in this chapter) have more recently come to the forefront of technology development, thus creating new challenges for regulators, civil aviation authorities, and air traffic management service providers. More will be discussed on the technology involved in those systems in a later chapter, but it is still important to briefly examine EUROCONTROL's research efforts that will play a significant role in how AAM/UAM operations will roll out.

EUROCONTROL sponsors and participates in numerous research efforts act across a broad spectrum of aviation concerns. Among them are three that directly impact UAS integration across Europe: AURA, CORUS, and CORUS XUAM.

Aura

The AURA project aims to ensure seamless and safe cooperation between U-space and air traffic management (ATM/ATC) by developing procedures and systems to support the needed interface between these two systems.

The project's goal is to manage high volumes of manned and unmanned traffic operating safely and concurrently in the same airspace through an accurate view of the traffic.

This research intends to inspire a new part of the U-space industry, boosting economic development, competitiveness, and innovation in public services. Inspections, logistics, disaster relief operations, drone-related manufacturing and maintenance, as well as unmanned traffic management (UTM) services will be among the first activities to benefit from the AURA project.

The project outcomes will facilitate the work of regulatory bodies, which are working on establishing the rules which will govern urban air mobility, and the development of technological standards and solutions shared by all manufacturers and operators.

To make this a reality, AURA partners will determine the flight information that UTM and ATM systems must exchange in order to guarantee safe air operations. This information will be used to develop new procedures and systems to support these operations. All operational concepts and systems developed as part of the AURA project will then be tested in several validation exercises at different European sites. These validation exercises will combine both real flights and high-performance simulations with real traffic to validate the project's solutions and procedures.

The project is developing two solutions: the first focuses on the immediate needs for collaboration between U-space and ATM/ATC (collaborative U-space-ATM interface), whereas the second solution focuses on future needs, including UAM, mixed and non-segregated manned/unmanned traffic.

EUROCONTROL's role in this effort, and others, is to contribute to the development of the "medium- to long-term concept for a collaborative ATM-U-space environment." This solution facilitates seamless operations of drones and manned aviation in non-segregated airspace.

CORUS (See Previous Discussion)

As previously noted, the Concept of Operations for European Unmanned Traffic Management (UTM) Systems (CORUS) project encompasses two years of exploratory research in order to adopt a harmonized approach to integrating drones into very low level (VLL) airspace. CORUS's task is to describe in detail how U-space should operate so as to allow the safe and socially acceptable use of drones.

CORUS's goal is to foster the uniform implementation of UTM in Europe, making cross-border operations easier and giving businesses with an interest in the sector the opportunity to provide services. The project will also consider the diverse needs of Europe's future U-space, balancing regional specificities while ensuring the safety of airspace users and the public.

The CORUS project needs to address two intrinsically linked issues:

1. Drones still do not operate as reliably as manned aircraft, and
2. Drones are not yet entirely accepted by society, especially when it comes to privacy concerns.

The project also faces competition from similar initiatives development and it challenges some of the basic assumptions in this rapidly changing environment.

The project is developing a U-space Concept of Operation (ConOps), through an iterative process. The CORUS project involves considerable consultation with stakeholders.

EUROCONTROL's role is to serve as the project's main contributor and coordinator.

Conclusion

The European Union and its Member States have engaged in intense efforts to integrate unmanned systems into their joint and national airspaces. They have worked closely with the FAA and NASA in the US, and are in many respects ahead of the game. Coordinated and harmonized regulations and standards are necessary to bring the whole concept of a U-space to fruition, and EASA and the EU have been leaders in that respect. ICAO's role in that development is the focus of the next chapter.

References

OJ L 152 (June 11, 2019). on the rules and procedures for the operation of unmanned aircraft. 45–71. http://data.europa.eu/eli/reg_impl/2019/947/oj.

https://www.easa.europa.eu/document-library/agency-decisions/ed-decision-2019021r.

https://www.easa.europa.eu/documentlibrary/easy-access-rules/online-publications/easy-access-rules-unmanned-aircraft-systems.

https://www.easa.europa.eu/document-library/acceptable-means-of-compliance-and-guidance-materials/amc-gm-commission-1.

"Specific Operations Risk Assessment" a process to create, evaluate and conduct an Unmanned Aircraft System operation, published at: http://jarus-rpas.org/sites/jarus-rpas.org/files/jar_doc_06_jarus_sora_v2.0.pdf.

https://www.easa.europa.eu/sites/default/files/dfu/amc_gm_to_commission_implementing_regulation_eu_2019-947_-_issue_1_amendment_1_0.pdf.

OJ L 139 (April 23, 2021a), 187–188. http://data.europa.eu/eli/reg_impl/2021/666/oj.

OJ L 139 (April 23, 2021b), 161–183. http://data.europa.eu/eli/reg_impl/2021/664/oj.

OJ L 139 (April 23, 2021c), 184–186. http://data.europa.eu/eli/reg_impl/2021/665/oj.

https://www.easa.europa.eu/sites/default/files/dfu/Opinion%20No%2001-2020.pdf.

https://corus-xuam.eu (copyrighted by EUROCONTROL Experimental Center 2020).

EUROCONTROL website (accessed September, 28 2021).

SJU, European ATM Master Plan – roadmap for the safe integration of drones into all classes of airspace. http://jarus-rpas.org/content/jar-doc-06-sora-package.

3

ICAO

The International Civil Aviation Organization (ICAO) is the global overseer of civil aviation (see the discussion in Chapter 1 for the history and scope of the organization). ICAO plays a major role in the development of regulations, standards, and recommended practices in all domains in civil aviation, and has taken particular interest in the development of unmanned aircraft systems and airspace integration since about 2015. ICAO has sponsored a number of programs and efforts to advance the field of knowledge of unmanned aircraft and facilitate the integration of UAS into international airspace, as well as within its 193 Member States. [Note: ICAO generally does not disclose publication dates for ICAO documents posted on their website, so the publications summarized in this chapter represent the most current versions publicly available as of the publication of this book.]

By way of guidance, ICAO has published several documents that are intended to inform the aviation community about the standards, recommended practices, model regulations, and Advisory Circulars dealing with the operation of RPAS:

- ICAO Model UAS Regulations
- UTM Guidance
- ICAO RPAS CONOPs
- ICAO U-AID Guidance
- UAS Toolkit

In addition, the organization sponsors panels, symposia, task forces, advisory groups, expert groups, training courses, and workshops, all intended to provide expertise, guidance, and resources for the entire aviation community.

This chapter will cover three of those publications: ICAO Model UAS Regulations; UTM Guidance; and ICAO RPAS CONOPS.

ICAO Model UAS Regulations

Member States asked ICAO to develop a regulatory framework for unmanned aircraft systems (UAS) that operate outside of the IFR International arena. ICAO reviewed the existing UAS regulations of many States to identify commonalities and best practices that would

UAS Integration into Civil Airspace: Policy, Regulations and Strategy, First Edition. Douglas M. Marshall.
© 2022 John Wiley & Sons Ltd. Published 2022 by John Wiley & Sons Ltd.

be consistent with the ICAO aviation framework and that could be implemented by a broad range of States. The outcomes of this activity are ICAO Model UAS Regulations titled Parts 101, 102, and 149.

The ICAO Model UAS Regulations and companion Advisory Circulars (ACs) offer a template for Member States to implement or to supplement their existing UAS regulations. These regulations and ACs are intended to be a living document and will evolve as the industry matures, providing States and regulators with internationally harmonized material based on the latest developments.

Part 101 highlights:

- All unmanned aircraft (UA) should be registered.
- UA weighing 25 kg or less and operating in Standard UA Operating Conditions (101.7) require no additional *operational* review; however, if the UA weighs more than 15 kg, the UA must be inspected and approved under 101.21 or 102.301.

Part 102 highlights:

- Addresses all UA operations using UA that weigh more than 25 kg or those weighing 25 kg or less but do not adhere to Part 101 requirements.
- Enables on-going operations or one-time events through certification.
- Enables a more expeditious review when manufacturers declare a type or model of UA as being sufficiently tested for a specific operational category or that has received an approval through an Approved Aviation Organization.

Part 149 highlights:

- Promotes the use of an Approved Aviation Organization to serve as a designee authorized by the CAA to perform specific tasks. Once the organization has been certified, the authorized tasks (remote pilot licensing, UA inspection, UA approval, etc.) may provide more expeditious processing and may reduce the workload for CAA Inspectors.

Advisory Circulars

The followings ACs have been provided for additional insight into the ICAO Model UAS Regulations:

AC 101-1: Provides guidance associated with rule 101 regarding unmanned aircraft system (UAS) operations in the Open Category.

AC 102-1: Provides guidance associated with rule 102 regarding the Specific Category, UAS authorizations or a UAS operator certificate (UOC). It also addresses requirements for manufacturers.

AC 102-37: Provides guidance for the carriage of dangerous goods transported by UA. This document is helpful to understand the risks and responsibilities for safe carriage and includes information for packing and marking.

DRAFT Canada AC 922-001, RPAS Safety Assurance

This draft Advisory Circular (AC) provides information for consideration by States to assist them with UAS regulations under development in setting standards for the manufacturer's Declaration of Compliance (DOC). While this draft AC is specific to RPAS advanced operations in Canada, this material can be studied by States to assist in the development of their State's individual Safety Assurance standards for manufacturers.

AC 922-001 is an example of performance-based criteria. Transport Canada specifies to what standard the manufacturer must comply for operations in controlled airspace, over people, or near people, and identifies classifications of injury severity. When the final version of this AC is published, the draft version will be replaced.

Humanitarian Guidance: This guidance material, located at https://www.icao.int/safety/UAID, relates to many UA operations that provide aid in locations during an emergency response as well as on-going humanitarian deliveries. The material includes helpful forms and an application process for expedited review by the Civil Aviation Authority (CAA).

As other best practice and regulatory materials are identified, they will be added to this website.

Because the standards setting process must take into consideration a State's overall UAS regulatory framework, ICAO will not be providing a safety assurance document for setting standards applicable to manufacturers (Part 102.301), cited in the ICAO Model UAS Regulations.

ICAO is exploring opportunities through which direct assistance may become available to support States in implementing UAS regulations, oversight, and training.

UTM Guidance

Unmanned Aircraft Systems Traffic Management (UTM) – A Common Framework with Core Principles for Global Harmonization Edition 3 is the latest iteration of ICAO's UTM activities.

To summarize:

ICAO recognizes ever increasing demands on Member States and regulators to be innovative and proactive when approving unmanned aircraft systems traffic management (UTM) proposals. Presumably negative impacts on safety, security, the environment, system reliability, and economic efficiency could result from insufficient international harmonization of regulations and standards.

ICAO envisions that civil aviation authorities and Air Navigation Service Providers (ANSPs) will implement UTM architectures to enable them to provide real-time information regarding airspace constraints and flight intentions to UAS operators and remote pilots, either directly or through a UTM service provider (USP). The responsibility for managing their operations safely within these constraints without receiving positive air traffic control services from the ANSP will fall on the UAS operator.

The UTM concept is already under development, but a common agreement or consensus on its framework and principles is an essential element towards ensuring global harmonization and system interoperability. ICAO is leading efforts by States, UAS industry leaders, academic institutions, and aviation professionals towards the development of this framework for UTM.

ICAO intends for this guidance document to provide those States that are considering the implementation of a UTM system with a framework and the core capabilities of a model UTM system. A UTM system must be capable of interacting with the existing air traffic management system in the short term and then to integrate with the ATM system in the long term. The safety and efficiency of the existing ATM system should not be negatively affected by the introduction and management, or integration, of unmanned traffic, nor by the development of an associated UTM infrastructure. A common framework will facilitate harmonization between UTM systems globally and will provide a phased approach towards integration into the ATM system. The end goal is to enable industry, manufacturers, service providers and end users to mature safely and efficiently without disrupting the existing manned aviation system.

Several components of a safe and effective UTM system have not been addressed in this version of the framework, such as design and certification standards of the UA, integration of UA operations in ATM, and potentially high-altitude airspace UTM systems. Future editions of this framework may address these issues, building on the foundation established by previous editions of the UTM Framework and information gathered by ICAO through the UTM request for information (RFI) process related to the DRONE ENABLE symposia. (The Drone Enable symposium is an annual event accessible on line at the https://www. icao.int/Meetings/DRONEENABLE4/Pages/default.aspx/drone website, a series begun in 2017 that is well worth watching to stay abreast of ICAO'S activities in the UTM/ATM/ Drone realm. The scheduled 2020 event was apparently cancelled due to the COVID-19 pandemic.)

The ICAO vision for UTM is the safe, economical, and efficient management of UAS operations through the provision of core facilities and a seamless set of services operating collaboratively with all parties and involving airborne and ground-based functions. Like ATM, a UTM system would provide cooperative integration of humans, information, technology, facilities and services, supported by air, ground, and/or satellite communications, navigation, and surveillance.

UTM systems should be interoperable and compatible with existing ATM systems so as to facilitate safe, efficient, and scalable operations. Although system-level requirements for UTM systems are still under development, core principles can be created that will guide their development. Many principles existing in the current ATM structure are also applicable to UTM services.

Numerous factors should be assessed when a State is considering issuing an operational approval for a UTM system. The key considerations for a safe UTM system should be:

- Registration and identification
- Communications and geo-awareness/geo-fencing
- Types of UA and their performance characteristics (including navigation capabilities and performance)
- Adequacy and complexity of the existing airspace structure

- Spectrum availability and suitability
- Nature of the operation
- Type and density of existing and anticipated traffic (manned and unmanned)
- Operational capacity of the UTM system including any airspace constraints
- Levels of and extent of automation capabilities in the UTM system and in the UAS
- Regulatory structure
- Meteorological considerations
- The requirement for all UA in the UTM airspace volume to be cooperative
- Detection/separation of non-cooperative UA
- Management of aeronautical information service (AIS)/aeronautical data and
- Geographic information systems (GIS) data/additional geospatial data applicable to the UTM airspace

The safe operation of UAS (particularly BVLOS operations) in a UTM system will depend on a range of supportive and enabling services. UTM systems are imagined to provide some of these elements, but they will require enabling policy and regulatory frameworks that take into account evolving technological solutions. States can address many of these requirements as they prepare for implementation of a UTM system. The document provides a notional list of services that may be included, among other things, in a UTM architecture:

- Activity reporting service: a service that provides on-demand, periodic, or event-driven information on UTM operations occurring within the subscribed airspace volume and time (e.g. density reports, intent information as well as status and monitoring information). Additional filtering may be performed as part of the service.
- AIS: a service that enables the flow of aeronautical information/data necessary for the safety, efficiency, economy, and regularity of, in this case, UAS operations.
- Airspace authorization service: a service that provides airspace authorization from the delegated State authority to the UAS operator.
- Discovery service: a service that provides users of the UTM system with information on relevant services of varying levels of capability in a specific geographical volume of airspace (e.g. suppliers of meteorological information).
- Mapping service: a service that provides terrain and obstacle data (e.g. GIS) appropriate and necessary for meeting the safety and mission needs of individual UAS operations or for supporting UTM system needs for the provision of separation or flight planning services.
- Registration service: a service that enables UAS operators to register their UA and provide any required data related to their UAS. The system should also include a query function enabling authorized stakeholders (e.g. regulators or police services) to request registration data. See Appendix A for additional information.
- Restriction management service: a service that manages and disseminates directives (e.g. safety bulletins) and operational and airspace restrictions from the CAA or ANSP to UAS operators and remote pilots, including in the form of NOTAMs.
- Flight planning service: a service that, prior to flight, arranges and optimizes intended operational volumes, routes and trajectories for safety, dynamic airspace management, airspace restrictions and mission needs (this is not intended to refer to the existing manned aircraft flight planning services).

- Conflict management and separation service (refer to Doc 9854 – *Global Air Traffic Management Operational Concept*), including:
  - Strategic deconfliction service: a service consisting of the arrangement, negotiation, and prioritization of intended operational volumes, routes, or trajectories of UAS operations to minimize the likelihood of airborne conflicts between operations.
- Tactical separation with manned aircraft service: a service that provides real-time information about manned aircraft so that UA remain well clear of manned aircraft.
- Conflict advisory and alert service: a service that provides remote pilots with real-time alerting through suggestive or directive information on UA proximity to other airspace users (manned or unmanned).
- Conformance monitoring service: a service that provides real-time monitoring and alerting of non-conformance to intended operational volumes, routes, or trajectories for a UAS operator or remote pilot.
- Dynamic reroute service: a real-time service that provides modifications to intended operational volumes, routes, or trajectories to minimize the likelihood of airborne conflicts and maximize the likelihood of conforming to airspace restrictions, while enabling completion of the planned flight. This service would include the arrangement, negotiation, and prioritization of in-flight operational volumes, routes, or trajectories of UA operations while the UA is airborne.
- Identification service: a service that makes it possible to identify an individual UA and the associated nationality and registration information. See Appendix A for additional information.
- Tracking and location service: a service that provides information to the UAS operator and the UTM system about the exact location of UA, in real time. See Appendix A for additional information.
- Meteorological service: a service that provides individual UAS operators/remote pilots or other UTM services with the meteorological information necessary for the performance of their respective functions.

The ideal UTM system could be thought of as a collection of services intended to ensure safe and efficient operations of UA within the UTM-authorized volume of airspace and is in compliance with regulatory requirements. It is anticipated that UAS operations could occur in both uncontrolled and controlled airspace, with each class of airspace potentially requiring specific services. When UAS operations occur in controlled airspace, UAS operators and/or the remote pilot would be required to follow the standard procedures and requirements for that airspace, unless an exemption or alternate procedure has been established, exempting those operating in the UTM system from the established airspace rules.

While this document does not specify technologies associated with these services, it does suggest various types of services. Ideally, a mature UTM system would provide for:

- Continued safety of all air traffic, manned and unmanned
- Safety of persons on the ground
- Complex low-level UA operations
- Ongoing support of technological advancements
- Evaluation of security and environmental risks
- Provision for a global, harmonized framework for low-level UTM

The safe and efficient integration of UAS, particularly small UA into existing airspace, presents numerous challenges. Indeed, recent studies forecast significant growth of UAS operations, leading to a shift of focus to operations in the low-level environment and over populated areas, with a variety of types of operations and UA.

The document also includes a discussion of the numerous gaps, issues, and challenges that must be addressed to enable safe UAS operations within the UTM system.

The framework outlined in this document is not intended to endorse or propose any specific UTM system design or technical solutions to the UTM challenge, but rather to provide an overarching framework for such a system. "The intent is for this to be a living document and as new or additional information is gained, the UTM framework will be updated."

ICAO recognizes that the developmental nature of the UTM concept makes it difficult to predict how a follow-on framework would be organized, validated, and certified. Continued participation from industry and/or future business advocates is necessary to explore the minimal set of safety issues in product deployment/development that has the potential to achieve global interoperability.

ICAO/RPAS Concept of Operations

This document contains the following disclaimer: "*This document is an unedited version of an ICAO publication and has not yet been approved in final form. As its content may still be supplemented, removed, or otherwise modified during the editing process, ICAO shall not be responsible whatsoever for any costs or liabilities incurred as a result of its use.*" Although not so stated, the document was prepared in March of 2017 and has not been updated since.

The Introduction refers the reader back to the Convention on International Civil Aviation (Doc. 7300), wherein the term "pilotless aircraft" is used (although not stated, it appears in Article 8) to prohibit such aircraft from being flown over the territory of another nation without that nation's approval. Actually, the term "pilotless aircraft" is not used in Article 8, but "aircraft capable of being flown without a pilot" is. The preferred term now is remotely piloted aircraft system (RPAS).

The purpose of the document is to "describe the operational environment of manned and unmanned aircraft, thereby ensuring a common understanding of the challenges and how the subsets that are remotely piloted can be expected to be accommodated and ultimately integrated into the airspace for international instrument flight rules (IFR) operations."

"ICAO will use this CONOPS to inform the Air Navigation Commission, States and ICAO expert groups to scope proposed amendments to ICAO Standards and Recommended Practices (SARPs) and Procedures for Air Navigation Services (PANS). As such, this document serves as a general framework to represent the perspective of stakeholders from Member States, including regulators, operators, airspace users, manufacturers, air navigation service providers (ANSPs) and aerodrome operators."

The problem statement describes a number of issues that the document is intended to address. First, it is necessary to clarify or understand what is meant by the term "RPAS." Is it the entire system (RPA, plus the remote pilot station (RPS), plus command and control (C2) link, plus other components), or is it confined to just a portion of that system, i.e. the

"RPAS operator conducts RPAS operations, however RPA are separated from other aircraft in the airspace?"

The introduction of beyond visual line of sight (BVLOS) operations into airspace occupied by manned aircraft is a major challenge. Consequently, ICAO identifies a need to establish the principles for RPAS operations in all classes of non-segregated airspace. All aspects of the air navigation system (regulations, procedures, initial and recurrent pilot training, etc.) combine to manage the safe, efficient flow of air traffic. The broad range of new capabilities distinguishing remotely piloted aircraft may compel modification of infrastructure, procedures, and policies to support the new technology demands. Integration of RPAS into unsegregated airspace must not create an undue burden on current airspace users and service providers and cannot compromise safety of the overall system.

Additional challenges are presented by the requirements of manned aviation that the pilot should "see and avoid" and "remain well clear" of other aircraft or dangerous situations, and communicate with air traffic control, when necessary, all of which affect the entire system. Innovative ATC communication architectures, traffic management procedures, airworthiness approvals of technical capability, the potential use of third-party communication service providers, and changes in the regulatory approvals and oversight regimes will all affect the traffic management system, and must be accounted for with updates and amendments to SARPS and PANS that currently only apply to manned aviation. The detect and avoid (DAA) capability may become a technical and regulatory necessity to ensure compliance with the see and avoid requirement for manned aviation. Thus, some alternate means of compliance (AMC) for RPAS must be included, where appropriate, in future versions of these documents.

ICAO forecasts that, by 2030, a large number of RPA will share the airspace with manned aviation, and some will be operating under IFR. Some RPAS operations may be conducted in accordance with IFR for a portion of their flight, while others will operate only under VFR. It is expected that RPA will operate in and transit through national and international airspace, as well as controlled and uncontrolled airspace. RPAs may depart from less congested aerodromes and arrive at similar conditions at their destinations, while others may use congested aerodromes. All RPA will be required to comply with the applicable procedures and airspace requirements defined by the Member State, including emergency and contingency procedures, which should be established and coordinated with the respective ANSPs. Accordingly, these types of RPA operations could include flying in both national and international airspace. In contrast, other RPA will only operate at low altitudes, where manned aviation activities are limited. Operations such as border protection, environmental uses, and wildfire and utility inspections could still mean transiting international airspace.

The scope of this CONOPS is currently limited to certificated RPAS operating internationally within controlled airspace under instrument flight rules (IFR) in non-segregated airspace and at aerodromes in a time frame beyond 2031. The scope does not consider fully autonomous aircraft and operations, visual line-of-sight (VLOS) operations, very low altitude airspace operations, and very high-altitude operations (above FL600) or carriage of persons and domestic operations. [Author's note: This would necessarily exclude the evolving Advanced Air Mobility and Urban Air Mobility concepts.] Generally, this section focuses on three areas of concern: RPAS operations, RPAS technology aspects, and airspace aspects.

"The (2030) time frame aligns with ICAO's Aviation System Block Upgrade (ASBU) estimated completion date for RPAS Block 3 activities. Block 3 represents a period when RPAS certification processes will be complete; avionics and ground systems made available; and State policies, regulations, procedures and guidance permitting routine and safe operations are in place." A footnote explains the ASBU reference: "(ICAO's ASBUs provide a global systems engineering approach to facilitate the advancement of air navigation and enable global harmonization, increased capacities and improved environmental efficiency. The ASBU framework is presented in the ICAO Global Air Navigation Plan (GANP) and provides broadly-defined objectives. The framework has four Blocks (0, 1, 2, and 3), each defining associated modules, objectives and timelines. Three RPAS modules are defined in Blocks 1 (2019), Block 2 (2025) and Block 3 (2031 onwards.)" These modules support the phased approach to airspace integration described in the UTM Guidance discussion above.

> "Future iterations of this document may expand this scope where evidence indicates unanticipated needs resulting from market growth, technology advances or other unforeseen conditions. This expansion is likely to include operations at very high altitude, very low altitude, autonomous aircraft and remote pilot control of multiple RPA as these activities are already being actively pursued by operators."

Key assumptions underlying the approach to airspace integration are assumptions that can be applied to all RPAS operations and are described as follows:

1) Access to the airspace remains available to all, providing each RPA is capable of meeting pertinent conditions, regulations, processes, and equipage defined for that airspace.
2) New types of operations may need additional or alternative considerations, conditions, regulations, processes and operating procedures; the objective should be to add only the minimum necessary to achieve safe operation.
3) The RPA has the functional capability to meet the established normal and contingency operating procedures for the class of airspace, aerodrome, etc., when such procedures are available.
4) The flight operation does not impede or impair other airspace users, service providers (such as air traffic management (ATM), aerodromes, etc.), or the safety of third parties on the ground and their property, etc.
5) The RPA must operate in accordance with Annex 2 – *Rules of the Air*.
6) The RPAS must meet the applicable certification/registration/approval requirements.
7) The operator must meet the applicable certification/approval requirements.
8) The remote pilot must be competent, licensed, and capable of discharging the responsibility for safe flight.

The general approach taken in this CONOPS is conformance with the existing aviation system and its planned evolution and updates, rather than significant modification or evolution. Some level of change is anticipated, but the expectation is that RPAS will be handled by ANSPs in the same manner as manned aircraft.

Three phases of RPAS accommodation are proposed:

- ***Accommodation vs. integration*** (allowing for early RPA flights on a temporary and transitional basis and in limited numbers before the required technology, standards, and regulations are in place vs. a future time in which RPA may be expected to enter the airspace system routinely without requiring special provisions).
- ***Integration from the present to 2025*** (using specialized authorizing techniques).
- ***Integration from 2025 onwards*** (a complete set of technologies, standards, regulations, guidance and procedures will not be available to support transparent integration across the wide array of RPAS and types of possible operations until 2031, the year that corresponds with ICAO ASBU Block 3 objectives for the RPAS module).

Subsets of UAS are RPAS, autonomous aircraft, and model aircraft. The term "UAS" includes all aircraft flown without a pilot on board that operate as part of a larger system. Autonomous aircraft differ from RPAS in that they do not permit intervention of a human pilot to achieve their intended flight. "Model aircraft are distinguished by their recreational use." [Author's note: This distinction is not so clear. While there may be "autonomous" model aircraft, they may not be in conformance with the rules and bylaws of the Academy of Model Aeronautics, for example. AMA's GUIDANCE FOR ADVANCED FLIGHT SYSTEMS (Failsafe, Stabilization, and Autopilot) states that "[m]odels using advanced flight systems allowing for automated or flight are permitted by AMA, provided the pilot remains in direct control and flies within visual line of sight. In such operations, a modeler must be able to override the automated and programmed features at all times. The specific automated functions allowed for this type of model operation are listed in this guidance document." Thus, an autonomous model aircraft, one that doesn't permit human intervention in the operation, when put to "recreational" use, may not qualify for the regulatory or community-base-organization definitions of a "model aircraft."]

RPAS component designs that potentially may require approval and oversight are listed below:

- Remotely piloted aircraft
- Remote pilot station
- C2 link
- Radio line-of-sight and beyond radio line-of-sight
- C2 link performance
- C2 link protection
- Third party C2 link service provision
- C2 link frequency spectrum and management
- Operational safety systems
- Detect and Avoid Capability
- RPAS/ATC communications
- System interfaces
- Human performance
- Automation and human intervention
- Categorization of RPA

- Airworthiness
- Special airworthiness considerations
  - RPAS classifications
  - Airworthiness and C2 Link service providers
  - Airworthiness approval and oversight
- RPAS operations
  - Aerodrome surface operations
  - RPA landing
  - Future operations
- Operators
  - Safety management
- Flight trajectories
  - Point-to-point trajectories
  - Defined trajectories
  - Dynamic trajectories
- Operational planning
  - RPAS operators
  - Air traffic management
  - Flight planning
- Special considerations
  - Defining and managing international flight
  - Delegated separation
  - In-flight handover between remote pilot stations (RPS)
  - In-flight transfer of C2 Link service providers
  - In-flight transfer of operators
  - Emergency and contingency operations
  - Flight data recording
- Personnel licensing
  - Remote pilot
  - General remote pilot licensing provisions
  - Credit for prior experience
  - Class and type ratings
  - RPA category
- Remote pilot instructors
- Remote pilot license (RPL) examiners
- RPAS maintenance personnel
- Air traffic controllers
- Operating environments
  - International airspace rules and procedures
  - Airspace requirements and RPAS capabilities
- RPAS performance limitations

In summary, these elements of the current and future aviation system that will or must accommodate or integrate RPAS into those systems, as envisioned by ICAO, will provide the basis for the structure of the penultimate chapter of this book.

ICAO U-AID Guidance

Humanitarian aid and emergency response operations, hereinafter referred to collectively as "U-AID," can include scheduled and unscheduled medical deliveries or provide emergency response to victims of natural or man-made disasters.

ICAO is part of an international effort to identify and streamline emergency preparedness at international airports. This is done in partnership with the United Nations Office for the Coordination of Humanitarian Affairs (OCHA) and additional aviation and humanitarian partners. ICAO is also in a collaborative arrangement with the World Health Organization (WHO) for the prevention and management of Public Health events in civil aviation (CAPSCA). These efforts assist member States and align with the United Nations sustainable development goals.

This guidance material is a resource for member States to enable humanitarian aid and emergency response operations using UAS and to enable an expedited review process for urgent operations. It is applicable for States that have already implemented UAS regulations and for States who are in the beginning stages of promulgating UAS regulations.

The document begins with an operational overview defining the types of humanitarian missions that could be covered by the guidance material, and provides on-line resources for the application and qualification process for real-time or future operations.

The guidance then moves on to cover specific areas of operational concern:

- Airspace rules (with specific reference to the ICAO UTM Framework).
- Environmental conditions (specifications for manufacturers that design and produce UAS).
- Contingency and emergency operational status (such as loss of the command-and-control link, failure or degradation of the detect and avoid system, loss of airborne radio communications, and emergency landings, requiring planning and specific procedures).
- System Specifications (Air, Ground, Cyber), including software, hardware, communications, frequency spectrum and emergency recovery and contingency.
- Crew and personnel training and descriptions (including a checklist of minimum requirements for remote crew and personnel competency).

A second major topic is the requirements for the transport of goods for humanitarian aid and emergency response, some of which may be considered "dangerous goods" that fall under specific regulatory scrutiny (goods that have one or more inherent hazardous characteristics). The scope of this guidance is for circumstances where a State has determined that the use of UAS to transport dangerous goods for humanitarian aid and emergency response is appropriate. The basic regulations are found in Annex 18 to the Convention on International Civil Aviation, *The Safe Transport of Dangerous Goods by Air* and *Technical Instructions for the Safe Transport of Dangerous Goods by Air* (Technical Instructions, Doc. 9284).

The document defines "dangerous goods," describes situations where transport of dangerous goods by an unmanned aircraft may be appropriate, and outlines operating procedures for such operations by reference to *The Safety Management Manual* (SMM) (Doc. 9859, 4th edition) that contains general guidance on implementation of Convention Annex 19 – Safety

Management, including the conduct of safety risk assessments. Also referenced is a new manual entitled *Guidance for Safe Operations Involving Aeroplane Cargo Compartments* (Doc. 10102), which provides guidance on specific safety risk assessments on the transport of items, including dangerous goods, in the cargo compartments of an airplane, which may be informative for UAS operations.

The guidance goes on to discuss "Safety Risk Management: Understanding Risk Assessment, Responsibility, and Mitigation," specifically, operational risk assessment for CAA authorization, the safety risk assessment process, and safety risk mitigations strategies. The safety risk assessment process section offers the standard risk assessment matrix found in many government publications dealing with risk assessment, with simple explanations for the assumptions and components. The document also provides links to the risk assessment processes developed by ICAO, JARUS, ASTM, the FAA, and a fifth website link that is no longer available. (All of these resources but the last are discussed elsewhere in this book.)

Appendix 1 explains classes and divisions of dangerous goods; Appendix 2 offers examples of dangerous goods that may be necessary for humanitarian aid or emergency response; Appendix 3 lists elements that should be included in a UA operator's policy and procedures manual for the safe transport of dangerous goods; Appendix 4 describes elements to consider as part of the UA operator's safety risk management procedures; and Appendix 5 provides references and links to ICAO, the International Air Transport Association (IATA), the United Nations, the World Health Organization, and the US Department of Health and Human Services websites that also provide guidance and regulatory treatment of and for the carriage of hazardous goods by air transport.

The ICAO/UAS Toolkit

This is an on-line tool intended for use for humanitarian efforts (search and rescue, firefighting, infrastructure monitoring, and research and development); air work, including all forms of photography and video surveillance; and carrying loads or discharging substances (e.g. crop dusting, insect control).

The second section on UAS regulations addresses the fact that UA operations will involve stakeholders familiar with aviation, as well as many who are not, referencing as a baseline the Chicago Convention (Doc. 7300). These stakeholders should be engaged from the beginning when developing the UAS regulations. Their early involvement will ensure that the regulations address the needs of these groups, while also educating them on expectations and feasibility. A summary of the regulatory development process includes discussions of the challenges facing the States that are developing UAS regulations; risk management efforts to integrate risk analysis and assessment into Civil Aviation Authorities planning and decision-making processes; the incremental introduction of regulations; the concept of performance-based or prescriptive regulations; the scope of regulatory work; how a risk-based approach works in the context of regulatory activates activities, such as complexity factors, including the size of the aircraft, location, altitudes, airspace classification and complexity of the operation, day/night operations and mitigations that may be imposed; provides sample components of a categorization scheme; and, finally, a recommended assessment of operational risk by the operator.

The third section of the Toolkit deals with training and education of general recommendations, referencing the RPAS Manual (Doc. 10019). It offers guidance on how CAA personnel should be educated; competence criteria for operators and remote pilots; remote pilot minimum age recommendations; professional remote pilot training and competencies; theoretical training learning objectives; practical skills training; practical test recommendations; training facility approval criteria; medical standards for remote pilots; land license recognition or credit by States.

The fourth section goes into additional considerations, such as a "whole government approach," the importance of stakeholder consultations, collaboration with civil aviation authorities, public outreach for education and awareness, safety campaigns, enforcement, incident and accident reporting, privacy, and UAS dedicated test facilities and airspace.

The last narrative section covers the authorization process and the applicability of Annex 2, Rules of the Air.

Also included in the Toolkit are helpful links for recreational operators, access to State regulations, and downloads of forms for operators and regulators. In addition, the document offers rules and guidance for regulators in developing regulations and policies.

RPAS Manual (Doc.) 10019 1ˢᵗ Edition

The last publication relevant to our discussion of ICAO's role in the development of UAS airspace integration is the RPAS Manual, published in January of 2015. This 166-page guidance document is probably due for an update or revision due to the many developments in the technology since 2015. The forward to the Manual states, in part:

> "The goal of ICAO in addressing RPAS is to provide an international regulatory framework through Standards and Recommended Practices (SARPs), with supporting Procedures for Air Navigation Services (PANS) and guidance material, to underpin routine operation of RPAS throughout the world in a safe, harmonized and seamless manner comparable to that of manned operations. Most importantly, introduction of remotely piloted aircraft into non-segregated airspace and at aerodromes should in no way increase safety risks to manned aircraft."

> "The content of this manual was developed over a period of three years with input from many groups of experts from RPAS inspectors, operators and manufacturers, pilot representatives, air navigation service providers (ANSPs), air traffic control representatives, accident investigation bureaus, human performance specialists, surveillance and communications experts and others. It is based upon the latest forms of technology available at the time of its publication. As such, it will be subject to a regular revision process that will be based on development of SARPs and PANS and input from the RPAS community."

As noted above, the manual has not been updated since 2015, so the many developments in UTM/U-space and AAM/UAM are not included in the content.

The manual is divided into 15 chapters dealing with all aspects of UAS technology and operations.

Chapter 1 is "ICAO regulatory framework and scope of the manual." The chapter includes the history and foundations of the legal framework, the purpose and scope of the manual, and guiding principles or considerations.

Chapter 2, "Introduction to RPAS," offers an overview and description of RPA, their components and operations.

Chapter 3, "Special authorizations," provides an overview and summary of Annex 2 general operating rules.

Chapter 4, "Type certification and airworthiness approvals," goes into details of the process for obtaining airworthiness certificates for RPAS and their components, operations, continuing airworthiness, validation, and responsibilities of States in that process.

Chapter 5 covers RPA registration (nationality and registration marks).

Chapter 6 outlines the "Responsibilities of the RPAS operator," including personnel, oversight, document retention, communication service providers, and operating facilities.

Chapter 7 "Safety management," discusses state safety programs, SMS processes for RPAS operators, and hazard identification and risk assessment.

Chapter 8 sets forth "Licensing requirements" and guidance for the regulators, licensing authorities, the RPAS instructor and observer's competency, and medical assessments.

Chapter 9 covers the various aspects of RPAS Operations.

Chapter 10 goes into great detail about the detect and avoid (DAA) technology requirements and capabilities.

Chapter 11 discusses the Command and Control (C2) link (architecture, management, protection, lost link, and recovery).

Chapter 12 addresses ATC communications (voice and data link, VLOS requirements, required communications performance, and minimum communications airborne equipment).

Chapter 13 covers Remote Pilot Stations and their configurations, display and control requirements, multiple UAS operations, and human factors considerations.

Chapter 14 is devoted to Integration of RPAS into ATM and ATM procedures. The discussion includes integration principles, flight rules, and ANSP SMS.

Chapter 15 explores the use of aerodromes, the application of Annex 14, aerodrome integration issues, controlled aerodrome environment, and flight information services.

Returning to Chapter 14, the scope of the chapter does not include operations in segregated airspace, nor to ground operations. Certain integration principles are offered:

- The advances and development of associated procedures. The process begins with limited access to airspace, and while some RPA may eventually be able to seamlessly integrate with manned flights, many may not.
- When adding any new type of airspace user into the existing air navigation system, consideration must be given to minimizing risk to all airspace users. States and service providers under oversight should therefore apply safety management principles and analyses to the introduction of RPAS operations. These principles and analyses should reflect on-going developments in RPAS capabilities.

- RPAS operations should conform to the existing airspace requirements. These airspace requirements include, but are not limited to, communication, navigation and surveillance requirements, separation from traffic, and distances from clouds.
- *Controlled airspace.* In order for RPA to be integrated into non-segregated controlled airspace, the RPA must be able to comply with existing ATM procedures. In the event that full compliance is not possible, new ATM procedures should be considered by the aviation authorities and/or ANSPs in consultation with the RPAS operator and representatives of other airspace user groups. Any new ATM procedures should be kept as consistent as possible with those for manned flights to minimize disruption of the ATM system.
- *Uncontrolled airspace.* In order for RPA to be integrated into non-segregated uncontrolled airspace, the RPA will need to be able to interact with other airspace users, without impacting the safety or efficiency of existing flight operations.

The principle enunciated in the "controlled airspace" paragraph anticipates, in a way, the emergence of U-space (or UTM) concepts, which were just beginning to draw attention in the year that this manual was published. The list of issues arising from more recently advanced airspace integration developments are well represented in this document. They are:

- Airspace requirements
- Take-off and landing phases
- En-route phase
- VFR
- IFR
- Communication, navigation, and surveillance (CNS)
- Transponder operations
- RPAS unique procedures
- Flight rules/Right of way
- RPAS performance requirements
- ATM procedures
- Flight plans
- Controller training
- ANSP SMS
- Traffic complexity
- Latency of RPA response
- Conspicuity
- Non-standard method of communication
- RPA sensitive to hazardous meteorological conditions
- Acceptance by airspace users and ATCOs

ICAO hosted its Drone Enable 2021 Symposium in April of 2021. The presentations in the 5-day event included:

- The RPAS Panel session provided an overview of the work being undertaken by ICAO and the progress being made with RPAS regulatory materials.

- The UAS ADVISORY GROUP (UAS-AG) summarized the work of the Group, including the most recent updates to the UTM Framework based on the information derived from DRONE ENABLE/3, updates to the ICAO UAS Toolkit, and the results of the DRONE ENABLE 2021 RFI Process.
- The ICAO MODEL UAS REGULATIONS session provided an overview of the ICAO UAS Model Regulations developed to support UAS regulatory harmonization. The companion supporting guidance material was also discussed.
- ICAO AIRCRAFT REGISTRATION NETWORK UPDATE/DEMO discussed ICAO's work with the Korean Ministry of Land, Infrastructure and Transport on an application that will allow aircraft registry and other applications to interact with the Commercial Aviation Safety Team (CAST)/ICAO Aircraft Taxonomy. A prototype of the application was presented that demonstrated how such a system can be integrated using open-source code and API calls. The system allows consulting the CAST/ICAO taxonomy in real-time to request new aircraft codes when one is not available.
- Facilitators led open discussions on UTM System Certification Requirements.
- A presentation on UTM Integration into Aerodrome Environments/Activities.
- The CYBER RESILIENCE panel discussed the evolution of the aviation system, considering its needs for secure and resilient connections for exchange of safety critical messages and how an international aviation trust framework can help new entrants to be safely integrated in a non-segregated airspace, considering possible cyber threats. They discussed the potential threats in a digitally connected environment and the need for global regulations to guarantee acceptable levels of safety in UAS and UTM operations. (See Chapter 8 of this book for a more detailed treatment of this topic).
- BRAZILIAN FOCUS. This panel highlighted the entire process related to UAS operations in Brazil and involved all the key stakeholders, including operators and Brazilian Authorities. Attendees were provided with an understanding of the main goals and the challenges during the various processes. In addition, the panelists discussed the necessity of implementing a viable UAS Traffic Management system as a key enabler for UAS scalability.
- The ADVANCED/URBAN AIR MOBILITY panel of leaders in the field of advanced/urban air mobility discussed their recent advancements, the regulatory/technical challenges that remain, and what is needed to see this capability deploy safely.
- The FLIGHT RULES IN AN EVOLVING ENVIRONMENT panel explored future regulatory provisions that must include relevant flight rules that acknowledge the change in the role of the human with higher levels of automation. Given the time it takes to formulate, assess, and implement new provisions in the aviation system, it is critical to consider the new use cases for unmanned aircraft and advanced/urban air mobility aircraft and the provisions that would be required to ensure safe operations. How will the role of the human-in-the-loop affect the definition and implementation of flight rules and airspace management – for new entrants and for conventional aviation? Do we need to define a new set of flight rules or can we revise or expand the existing visual and instrument flight rules to encompass these new aircraft and operations?

Conclusion

In summary, ICAO offers a wide variety of publications, services, and resources for all stakeholders in the commercial aviation domain, and have taken a leadership role in the development of regulations, standards, policies, and practices for the RPAS community. Their RPAS panel conducts annual symposia and other working groups within the organization continue to provide valuable support to the aviation community.

Reference

Academy of Model Aeronautics Safety Program Handbook, available at: https://www.modelaircraft.org/sites/default/files/documents/100.pdf.

4

UAS Airspace Integration in the United States

The US airspace integration effort can be traced back to 1981, with FAA Advisory Circular 91.57 (AC 91.57). This circular proposed a policy whereby model aircraft operated by hobbyists would be exempt from the relevant provisions of Part 91 of the Federal Aviation Regulations (Part 91 consists of the airspace/rules of the road portion of the FARs, similar to EASA's SERA and Article 2 "Rules of the Air" from the Chicago Convention). The FAA has historically considered model aircraft to be aircraft that fall within the statutory and regulatory definitions of an aircraft, as they are devices that are "invented, used, or designed to navigate or fly in the air" (Title 49 US Code § 40102 and 14 CFR 1.1). Consistent with the FAA's enforcement policy (compliance before enforcement), the agency's oversight of model aircraft was guided by the level of risk posed by these types of aircraft. AC 91.57 acknowledged that model aircraft could pose a hazard to full-scale aircraft and recommended a set of voluntary operating standards for hobbyists to follow to mitigate these safety risks. To avoid actual airspace integration with manned aviation, the operating standards included restrictions from operating over populated areas and limiting flights around or over people until the aircraft were flight tested and demonstrated to be airworthy. In addition, operations were to remain below 400 ft above the surface, were required to give way and avoid operating near or around manned aircraft, and to use observers to assist situational awareness. A weight limit of 55 lb became the standard definition of a small UAS (sUAS).

Subsequent to the issuance of AC 91.57, the FAA was not particularly concerned with unmanned aircraft until it issued a policy statement in 2007, entitled "US Department of Transportation, Unmanned Aircraft Operations in the National Airspace System, Docket No. FAA-2006-25714 (2007)." This was prompted in part by the recent availability of military-grade ISR (intelligence, surveillance, and reconnaissance) drones that were then in use in conflict zones around the world by US armed forces and its allies. In addition, small, recreational-type UAs fitted with cameras were being used by aerial photographers and agricultural interests for domestic commercial purposes. At the time, there was no specific federal aviation regulation that prohibited such activities, so the FAA relied upon its "compliance vs. enforcement" philosophy to discourage such uses until a regulatory structure could be put in place. This met with considerable resistance from entrepreneurs and commercial operators who maintained that the FAA had no Congressionally mandated regulatory authority to prevent commercial drone operations, in spite of a body of regulations that covered virtually every aspect of aviation except for unmanned aircraft, if that category could be said to fall outside of the definition of an "airplane." Those gaps in regulatory treatment of remotely piloted aircraft were eventually filled by the US Congress in a series of legislative responses to the demands for UAS regulations and standards.

In the context of this discussion there are two kinds of integration: into the airspace and into the regulatory system. The next section describes the second kind, as reflected by US federal statutes and regulatory responses thereto by the FAA. The activities described in the preceding paragraphs were prompted, in part, by lawmakers and regulators coming to the realization that unmanned aircraft are here to stay and that a regulatory structure to deal with the new technology was necessary to maintain safety in the NAS. As noted above, the FAA attempted to regulate UAS activities through policy statements and Advisory Circulars rather than specific rules. The rulemaking process in the US is long and cumbersome due to the requirements of the Administrative Procedures Act, which applies to all federal agencies. The FAA was put under considerable pressure from a variety of stakeholders to do something about the seemingly uncontrolled proliferation of small UAS, which had evolved from Styrofoam or balsa wood fixed-wing model aircraft to sophisticated, multisensor, multirotor, lightweight drones that could go places where a fixed-wing model aircraft could not. When the first Aviation Rulemaking Committee (ARC) met to draft a set of recommendations for UAS regulations in 2008, the camera-mounted multirotor drone was not even on the table for discussion, as the technology, for all intents and purposes, did not yet exist.

Consequently, the US Congress stepped in to fill the regulatory gaps with legislation specifically addressing unmanned aircraft.

The FAA Modernization and Reform Act of 2012 (Hereafter FMRA), Public Law 112-95, Title III – Safety, Subtitle B – Unmanned Aircraft Systems

Section 332 of the FMRA, entitled "Integration of civil unmanned aircraft systems into national airspace system," was the first act of Congress that explicitly mentioned unmanned aircraft systems and the intent of Congress to accelerate the integration of civil unmanned aircraft systems into the national airspace system. The Secretary of Transportation (the FAA is a subdivision of the Department of Transportation) was directed to develop a comprehensive plan for integration, to initiate the rulemaking process to define the acceptable standards for operation of civil UAS, to ensure that any civil UAS includes a sense and avoid capability, and to establish standards and requirements for the operator and pilot of a civil UAS, including standards and requirements for registration and licensing.

The Act mandated a phased-in approach to airspace integration, the creation of a safe airspace designation for cooperative manned and unmanned flight operations (the first hint of what would become UTM airspace), the establishment of a process to develop certification, flight standards and air traffic requirements for civil UAS operations at test ranges (provided for in another section of the Act), develop the best methods to ensure safe operations of civil unmanned aircraft systems and public unmanned aircraft systems simultaneously in the NAS, and, finally, to incorporate the plan into the annual NextGen Implementation Plan document.

In the US system, all aviation operations that are not military are either civil or public. Briefly stated, public aircraft are aircraft operated by federal or state and local governments (and any political subdivision of those governmental entities) exclusively for a governmental function, but not operated for compensation. The receipt of compensation would be considered a commercial use, and the flight would be ineligible for public aircraft operations (PAO) status. PAO status is determined on a flight-by-flight basis. The distinction has significant ramifications for the operator, in that civil operations must comply with all relevant Federal Aviation Regulations (FARs), whereas public aircraft only have to comply with Part 91 "General Operating and Flight Rules." No pilot or aircraft certification is required. The FAA's published policy is that it has no regulatory authority over public aircraft operations other than enforcing the airspace regulations. The issue of what is considered to be a "governmental function" is controversial, and there is no settled law that operators can rely upon for guidance. The statutes creating the public aircraft category (49 US C § 40102(a)(2), 40102(a)(41) and 40125(b)) are ambiguous on that point.

This brief digression is important, because under the current CONOPs for UTM (discussed below), public aircraft are only mentioned in the context of access to UTM operations data, but there is no discussion of the technical requirements for users of that airspace. Although UTM is an airspace management tool, as it is envisioned by NASA and the FAA, it also has many technical requirements, such as onboard equipment, registration, and remote ID, which fall under other sections of the FARs that the FAA says it cannot enforce with PAOs (Parts 21, 23, 25, 27, 29, 43, 45, 46, 47, 48, among others). As stated in the CONOPs for UTM, public safety entities, when authorized, can access UTM operations data as a means to ensure safety of the airspace and persons and property on the ground, security of airports and critical infrastructure, and privacy of the general public. Data can be accessed through dedicated portals or can be routed directly by service providers to public safety entities, local/tribal/state law enforcement agencies, and other relevant federal agencies (e.g. Department of Homeland Security (DHS)) on an as-needed basis. The general public can access data that is determined or required to be publicly available.

From the UTM CONOPs document:

> "Operations in UTM will need to rely on a similar construct through the issuance of Performance Authorizations. UAS operations are expected to vary greatly in CNS performance considering the many vehicle types and intended operations. The expectation is that the variance in performance will be managed by the USSs when providing different services. The USSs will need to account for the variance while maintaining safety and equity in the airspace.
>
> The regulatory side of the FAA – Flight Standards, Aircraft Certification, others as needed – will have a role in approval of UAS operations in UTM. Generally, it is not expected that the regulator will dictate specific CNS performance for airspace that is not under ATC control. These CNS performance requirements will either be a product of the Operator-specific safety case or determined more agilely by the USSs to effectively perform the offered services. It is also expected that there will be general principles of efficient airspace use."

It is thus unclear how the FAA will impose and enforce specific equipage requirements on public aircraft operations that may do more than access UTM operational data, but may well operate in the UTM airspace, constrained only by the operational requirements for use of the designated airspace. It is assumed that the FAA is working through the problem as it advances the UTM concept through evaluation and implementation.

Introducing a new technology, procedure or airspace rule into the US national airspace system requires a thorough safety analysis before the FAA can authorize the change. This has to be done through the rulemaking process, with publication in the Federal Register and opportunities for public comment and resolution. The safety analysis involves a lengthy process that includes a review of the current relevant regulations, Advisory Circulars, Special Rules, and policies to determine if the proposed change can comply with the existing regulations and standards. Exemptions and waivers can be granted through a formal petition process, which is again subject to public scrutiny. Changes to aviation products such as airspace designations and reporting points, IFR approach procedures, prohibition of flights in specified regions, airworthiness directives, petitions for exemptions to certain FARs, etc., are posted on the Federal Register, almost on a daily basis. Anyone with an opinion or input, no matter how helpful or unhelpful, can deliver a comment to the FAA that must be dealt with in some fashion.

The FAA's authority to issue rules regarding aviation safety is found in Title 49 of the United States Code. Subtitle I, Section 106 describes the authority of the FAA Administrator. Subtitle VII, Aviation Programs, describes in more detail the scope of the agency's authority. Part A, Subpart I, Section 40103 charges the FAA with prescribing regulations to assign the use of airspace necessary to ensure the safety of aircraft and efficient use of airspace.

Returning to PL 112-95 (commonly known as the "FMRA"), Subtitle B, Unmanned Aircraft Systems, Section 332 is entitled "Integration of Civil Unmanned Aircraft Systems into National Airspace System."

This section of the Act specifies and mandates that the FAA engage in certain activities and meet benchmarks of progress towards the full integration of unmanned aircraft into the NAS. The FAA failed to meet most of those benchmarks, but some progress has been made in spite of the delays.

In summary, the Act required the FAA to:

- Develop a comprehensive plan to safely accelerate the integration of civil unmanned aircraft systems into the national airspace system.
- Engage in the rulemaking process to:
  - Define the acceptable standards for operation and certification of civil unmanned aircraft systems;
  - Ensure that any civil unmanned aircraft system includes a sense and avoid capability;
  - Establish standards and requirements for the operator and pilot of a civil unmanned aircraft system, including standards and requirements for registration and licensing.
- Determine the best methods to enhance the technologies and subsystems necessary to achieve the safe and routine operation of civil unmanned aircraft systems in the national airspace system.

- Take a phased-in approach to the integration of civil unmanned aircraft systems into the national airspace system (and a timeline for that approach).
- Create a safe airspace designation for cooperative manned and unmanned flight operations in the national airspace system.
- Establish a process to develop certification, flight standards, and air traffic requirements for civil unmanned aircraft systems at test ranges where such systems are subject to testing.
- Determine the best methods to ensure the safe operation of civil unmanned aircraft systems and public unmanned aircraft systems simultaneously in the national airspace system.
- Incorporate the plan into the annual NextGen Implementation Plan document (or any successor document) of the Federal Aviation Administration.

Not later than one year after the date of enactment of the Act, the Secretary (of Transportation) was to approve and make available in print and on the Administration's internet website a five-year roadmap for the introduction of civil unmanned aircraft systems into the national airspace system, as coordinated by the Unmanned Aircraft Program Office of the Administration. The Secretary shall update the roadmap annually.

A final rule on small unmanned aircraft systems that allows for civil operation of such systems in the national airspace system, to the extent the systems do not meet the requirements for expedited operational authorization under Section 333 of the Act, was required within 18 months after the Comprehensive Plan was submitted to Congress. The Final Rule, based in part on the recommendations of the UAS ARC in April of 2009, was not released as a Notice of Proposed Rulemaking until 15 February 2015, and did not become effective until 29 August 2016 (a period of 7 years, 4 months, and 28 days).

Other provisions in the Act supporting airspace integration were the establishment of seven UAS test sites and creation of special permanent areas in the Arctic regions where small unmanned aircraft may operate 24 hours per day for research and commercial purposes.

More legislation dealing with unmanned aircraft followed.

The FAA Extension, Safety, and Security Act of 2016 Title II, Subtitle B-UAS Safety (Pub. L. 114-190)

This statute amended the FMRA and ordered the FAA to undertake additional tasks to, among other things, coordinate efforts with the Administrator of the National Aeronautics and Space Administration, and continue development of a research plan for unmanned aircraft systems traffic management development and deployment. In developing the research plan, the Administrator shall (A) identify research outcomes sought and (B) ensure the plan is consistent with existing regulatory and operational frameworks, and considers potential future regulatory and operational frameworks for unmanned aircraft systems in the national airspace system. The research plan shall include an assessment of the interoperability of a UTM system with existing and potential future air traffic management systems and processes.

The FAA Reauthorization Act of 2018 (Pub. L. 115-254)

This statute addressed a multitude of issues yet unresolved or fully examined by the FAA with regard to unmanned aircraft. Among the 43 sections devoted to UAS are the following that have direct application to UTM or airspace requirements.

Section 351(b) Applications: "The Secretary shall accept applications from State, Local, and Tribal governments, in partnership with unmanned aircraft system operators and other private-sector stakeholders, to test and evaluate the integration of civil and public UAS operations into the low-altitude national airspace system."

Section 351(c) Objectives: "The purpose of the pilot program is to accelerate existing UAS integration plans by working to solve technical, regulatory, and policy challenges, while enabling advanced UAS operations in select areas subject to ongoing safety oversight and cooperation between the Federal Government and applicable State, Local, or Tribal jurisdictions, in order to:

> (1) accelerate the safe integration of UAS into the NAS by testing and validating new concepts of beyond visual line of sight operations in a controlled environment, focusing on detect and avoid technologies, command and control links, navigation, weather, and human factors; and ...

> (4) identify the most effective models of balancing local and national interests in UAS integration."

Section 376 Plan for Full Operational Capability of Unmanned Aircraft Systems Traffic Management:

(a) In general: The Administrator, in coordination with the Administrator of the National Aeronautics and Space Administration, and in consultation with unmanned aircraft systems industry stakeholders, shall develop a plan to allow for the implementation of unmanned aircraft systems traffic management (UTM) services that expand operations beyond visual line of sight, have full operational capability, and ensure the safety and security of all aircraft.

(b) COMPLETION OF UTM SYSTEM PILOT PROGRAM. The Administrator shall ensure that the UTM system pilot program, as established in Section 2208 of the FAA Extension, Safety, and Security Act of 2016 (49 U.S.C. 40101 note), is conducted to meet the following objectives of a comprehensive UTM system by the conclusion of the pilot program:

> (1) In cooperation with the National Aeronautics and Space Administration and manned and unmanned aircraft industry stakeholders, allow testing of unmanned aircraft operations, of increasing volumes and density, in airspace above test ranges, as such term is defined in Section 44801 of title 49, United States Code, as well as other sites determined by the Administrator to be suitable for UTM testing, including those locations selected under the pilot program required in the October 25, 2017, Presidential Memorandum entitled, "Unmanned Aircraft Systems Integration Pilot Program" and described in 82 Federal Register 50301.

(2) Permit the testing of various remote identification and tracking technologies evaluated by the Unmanned Aircraft Systems Identification and Tracking Aviation Rulemaking Committee.

(3) Where the particular operational environment permits, permit blanket waiver authority to allow any unmanned aircraft approved by a UTM system pilot program selectee to be operated under conditions currently requiring a case-by-case waiver under part 107, title 14, Code of Federal Regulations, provided that any blanket waiver addresses risks to airborne objects as well as persons and property on the ground.

The mandated implementation plan's contents should:

(1) include the development of safety standards to permit, authorize, or allow the use of UTM services, which may include the demonstration and validation of such services at the test ranges, as defined in Section 44801 of title 49, United States Code, or other sites as authorized by the Administrator;

(2) outline the roles and responsibilities of industry and government in establishing UTM services that allow applicants to conduct commercial and non-commercial operations, recognizing the primary private sector role in the development and implementation of the Low Altitude Authorization and Notification Capability and future expanded UTM services;

(3) include an assessment of various components required for necessary risk reduction and mitigation in relation to the use of UTM services, including:

(A) remote identification of both cooperative and non-cooperative unmanned aircraft systems in the national air-space system;

(B) deconfliction of cooperative unmanned aircraft systems in the national air-space system by such services;

(C) the manner in which the Federal Aviation Administration will conduct oversight of UTM systems, including interfaces between UTM service providers and air traffic control;

(D) the need for additional technologies to detect cooperative and non-cooperative aircraft;

(E) collaboration and coordination with air traffic control, or management services and technologies to ensure the safety oversight of manned and unmanned aircraft, including:

(i) the Federal Aviation Administration responsibilities to collect and disseminate relevant data to UTM service providers and

(ii) data exchange protocols to share UAS operator intent, operational approvals, operational restraints, and other data necessary to ensure safety or security of the National Airspace System;

(F) the potential for UTM services to manage unmanned aircraft systems carrying either cargo, payload, or passengers, weighing more than 55 pounds, and operating at altitudes higher than 400 ft above ground level; and

(G) cybersecurity protections, data integrity, and national and homeland security benefits; and

(4) establish a process for:

(A) accepting applications for operation of UTM services in the national airspace system;

(B) setting the standards for independent private sector validation and verification that the standards for UTM services established pursuant to paragraph (1) enabling operations beyond visual line of sight, have been met by applicants; and

(C) notifying the applicant, not later than 120 days after the Administrator receives a complete application, with a written approval, disapproval, or request to modify the application.

Section 376(d) SAFETY STANDARDS. In developing the safety standards in subsection (c) (1), the Administrator:

(1) shall require that UTM services help ensure the safety of unmanned aircraft and other aircraft operations that occur primarily or exclusively in airspace 400 ft above ground level and below, including operations conducted under a waiver issued pursuant to subpart D of part 107 of title 14, Code of Federal Regulations;

(2) shall consider, as appropriate:

(A) protection of persons and property on the ground;

(B) remote identification and tracking of aircraft;

(C) collision avoidance with respect to obstacles and non-cooperative aircraft;

(D) deconfliction of cooperative aircraft and integration of other relevant airspace considerations;

(E) right of way rules, inclusive of UAS operations;

(F) safe and reliable coordination between air traffic control and other systems operated in the national airspace system;

(G) detection of non-cooperative aircraft;

(H) geographic and local factors including but not limited to terrain, buildings, and structures;

(I) aircraft equipage; and

(J) qualifications, if any, necessary to operate UTM services; and

(3) may establish temporary flight restrictions or other means available such as a certificate of waiver or authorization (COA) for demonstration and validation of UTM services.

(e) REVOCATION. The Administrator may revoke the permission, authorization, or approval for the operation of UTM services if the Administrator determines that the services or its operator are no longer in compliance with applicable safety standards.

(f) LOW-RISK AREAS. The Administrator shall establish expedited procedures for approval of UTM services operated in:

(1) airspace away from congested areas or

(2) other airspace above areas in which operations of unmanned aircraft pose low risk, as determined by the Administrator.

(g) CONSULTATION. In carrying out this section, the Administrator shall consult with other Federal agencies, as appropriate.

(h) SENSE OF CONGRESS. It is the sense of Congress that, in developing the safety standards for UTM services, the Federal Aviation Administration shall consider ongoing research and development efforts on UTM services conducted by:

(1) the National Aeronautics and Space Administration in partnership with industry stakeholders;

(2) the UTM System pilot program required by section 2208 of the FAA Extension, Safety, and Security Act of 2016 (49 U.S.C. 40101 note); and

(3) the participants in the pilot program required in the October 25, 2017, Presidential Memorandum entitled, "Unmanned Aircraft Systems Integration Pilot Program" and described in 82 Federal Register 50301.

(i) DEADLINE. Not later than 1 year after the date of conclusion of the UTM pilot program established in Section 2208 of the FAA Extension, Safety, and Security Act of 2016 (49 U.S.C. 40101 note), the Administrator shall:

(1) complete the plan required by subsection (a);

(2) submit the plan to:

(A) the Committee on Commerce, Science, and Transportation of the Senate and

(B) the Committee on Science, Space, and Technology and the Committee on Transportation and Infrastructure of the House of Representatives; and

(3) publish the plan on a publicly accessible Internet website of the Federal Aviation Administration.

Section 377 Early Implementation of Certain UTM Services

(a) IN GENERAL. Not later than 120 days after the date of the enactment of this Act, the Administrator shall, upon request of a UTM service provider, determine if certain UTM services may operate safely in the national airspace system before completion of the implementation plan required by Section 376.

(b) ASSESSMENT OF UTM SERVICES. In making the determination under subsection (a), the Administrator shall assess, at a minimum, whether the proposed UTM services, as a result of their operational capabilities, reliability, intended use, areas of operation, and the characteristics of the aircraft involved, will maintain the safety and efficiency of the national airspace system and address any identified risks to manned or unmanned aircraft and persons and property on the ground.

(c) REQUIREMENTS FOR SAFE OPERATION. If the Administrator determines that certain UTM services may operate safely in the national airspace system, the Administrator shall establish requirements for their safe operation in the national airspace system.

(d) EXPEDITED PROCEDURES. The Administrator shall provide expedited procedures for making the assessment and determinations under this section where the UTM services will be provided primarily or exclusively in airspace above areas in which the operation of unmanned aircraft poses low risk, including but not limited to croplands and areas other than congested areas.

(e) CONSULTATION. In carrying out this section, the Administrator shall consult with other Federal agencies, as appropriate.

(f) PREEXISTING UTM SERVICES APPROVALS. Nothing in this Act shall affect or delay approvals, waivers, or exemptions granted by the Administrator for UTM services already in existence or approved by the Administrator prior to the date of enactment of this Act, including approvals under the Low Altitude Authorization and Notification Capability.

The Response from the FAA and NASA

In 2011 the National Aeronautics and Space Administration (NASA) initiated its "UAS Integration into the NAS" project to support the aviation community, the FAA, and other government agencies and stakeholders by conducting research that would lead to development of standards and procedures that would allow safe integration of UAS into the US national airspace system (NAS). The research was focused on assisting industry by testing unmanned aircraft and electronic systems components to demonstrate how the technology could be safely operated in the NAS.

NASA worked with the FAA to define the protocols to create a flight test environment that employed real and simulated scenarios. This enabled industry to prove, or attempt to prove, through a series of actual (not simulated) flight tests, that any new technologies or procedures would work as intended. Specifically, NASA's research activities involved:

- **Technical Challenge-DAA: UAS Detect and Avoid Operational Concepts and Technologies**
  Develop Detect and Avoid (DAA) operational concepts and technologies in support of standards to enable a broad range of UAS that have Communication, Navigation, and Surveillance (CNS) capabilities consistent with IFR operations and are required to detect and avoid manned and unmanned air traffic.

- **Technical Challenge-C2: UAS Command and Control**
  Develop Satellite (Satcom) and Terrestrial-based Command and Control (C2) operational concepts and technologies in support of standards to enable the broad range of UAS that have Communication, Navigation, and Surveillance (CNS) capabilities consistent with IFR operations and are required to leverage allocated protected spectrum.
- **Demonstration Activity: Systems Integration and Operationalization (SIO)**
  Demonstrate robust UAS operations in the NAS by leveraging integrated DAA, C2, and state of the art vehicle technologies with a pathway towards certification to inform FAA UAS integration policies and operational procedures.

Testing these emerging technologies was among the first steps taken to demonstrate that the UA systems could be safely operated in any airspace, whether segregated or unsegregated. Phase 1 of the program began in 2011 and ended in 2018 and was focused on high altitude UAS operations (above 18 000 ft MSL). The concept of a special airspace designation to accommodate UAS traffic in unsegregated airspace was looming on the horizon in 2020 when NASA concluded the second phase of its test program, which concentrated on operations between 500 and 10 000 ft AGL. In that final phase, two companies partnered with NASA to participate in what was labeled a "Systems Integration and Operationalization Demonstration Activity" (SIO). The stated goal of the SIO was to advance the technology to support commercial UAS operations in the NAS. Towards that end, NASA partnered with American Aerospace Technology Incorporated (infrastructure inspections below 10 000 ft), Bell (critical medical transport in urban areas) and General Atomics Aeronautical Systems, Inc. (infrastructure inspections above 10 000 ft.) to conduct fight demonstrations in the NAS that simulate commercial missions. Concurrently, all three partners initiated individual programs to obtain FAA type certification for their respective demonstration aircraft.

The last SIO demonstration flight was conducted in February of 2021 (a simulated petroleum pipeline survey). All three demonstrations were declared a success by NASA and the FAA.

In late 2013 Amazon's founder Jeff Bezos announced in a television interview posted on YouTube that the company planned to use small drones to deliver packages to customer's doorsteps in an urban environment, and that he thought they could navigate the regulatory environment to launch such a service in "three to five years." Other large and well-funded companies soon followed by announcing their own versions of low altitude package deliveries.

The aviation community and the general public reacted negatively to the idea of a company with no aviation experience planning to introduce a new and untested air transportation mode to the world, and to do so by 2020. Thus, the concept of a UAS traffic management (UTM) system was born out of necessity. The technical challenges presented by visions of hundreds of small drones making thousands of flights every day over urban environments – like Amazon proposed – and to do it safely, were profound in 2013, and remain so to this day.

Among the many questions that have emerged out of this phenomenon are how to manage innumerable drones flying at low altitudes over a city; how to keep them from colliding with structures or one another; and how to handle emergency situations, especially if the pilots don't have the drones they are operating in visual line of sight.

It is unlikely that the FAA – or any other air navigation service provider in the world – could deal with this volume in the traditional manner, relying on the state of the art of air traffic management technology as it exists today. It would be impractical and fiscally impossible to hire, train, and equip enough air traffic controllers to perform these functions without overburdening the technical, operational, and human-based infrastructure that has successfully managed the US national (and international) airspace for over 60 years. That reality also does not begin to address the problem of designing and building a new air traffic management system that could operate within, alongside of, or integrated into the legacy ATM infrastructure.

As of 21 September 2021, 866,102 unmanned aircraft systems are registered to fly in the United States – and their numbers have increased exponentially since 2017, when the system recorded a "mere" 80,000 registered drones. Relatively few of those aircraft have an airworthiness certificate or can even qualify for one under the FAA's current regulatory scheme. These uncertificated aircraft cannot fly in the national airspace without waivers and exemptions, but rogue operators who choose to not go through the FAA approval process present a significant threat to aviation safety. Accordingly, there are many unanswered questions about how a dramatic change to the airspace will affect the overall safety of the system. The exponential growth of drones has motivated the United States Congress and the Federal Aviation Administration and other federal agencies in the US to become more aggressive in supporting research and development of what the US calls "UTM."

In response to these pressures, the NASA Ames Research Center in California initiated a research program in 2014 to test the potential for managing large numbers of drones flying at low altitude, integrating with other airspace users.

Sometime in 2015 or 2016 a NASA scientist, Dr. Parimal Kopardekar ("PK" to the rest of the world) revealed his vision for how to solve the problem, and obtained a patent for "Unmanned Aerial System Traffic Management to Enable Civilian Low Altitude Goods and Service Delivery by UAS" in 2019. The five-year project that led to the granting of this patent imagined an unmanned traffic management tool that would manage drone operations below 400 ft AGL. The concept eventually grew to contemplate operations in higher altitude airspace (Advanced Air Mobility – AAM).

In 2015 NASA sponsored a workshop and invited representatives from industry (Google, Amazon, and others), to discuss Dr. Kopardekar's ideas for a new kind of airspace structure. The positive response from the attendees provided NASA with the incentive to take the next steps.

In that same year NASA officially began the UTM project with "PK" as its first manager. To help get the ball rolling, the project hosted a three-day conference open to anyone interested in this new area of drone operations. More than 1200 people attended this "kick-off" event. The unexpectedly large attendance at that UTM conference sent a resounding message to all concerned about the future of unmanned aircraft and a new airspace configuration called "UTM." That message was to continue the work and get to the finish line sooner.

With the drone industry and the FAA now ready to move on the concept (see discussion of the FAA's congressional mandate above), NASA stepped in to lead and serve as a facilitator for the development of new UTM standards and procedures.

Similar to the U-space concept under development in Europe, NASA's notional UTM architecture is a "community-based traffic management system, where the operators and entities providing operation support services are responsible for the coordination, execution and management of operations," while complying with the rules and regulations established by the FAA.

The conceptual framework for UTM developed by NASA differs from the air traffic control system currently used by the Federal Aviation Administration to manage the nation's airspace. Like the regulatory environment for other air navigation service providers and national aviation authorities, the legacy system in the US largely depends on direct and real time communication between pilots and air traffic controllers when operating in controlled airspace (Classes A, B, C, D, and E). ATC clearances are required to operate in all those classes of airspace (although in Class E clearance is only required for IFR (instrument flight rules) and SVFR (special visual flight rules) operations). These are ICAO classifications, and each national aviation authority makes its own determination as to how it uses them. The existing ATC system depends upon a centralized provider of services (the FAA's Air Traffic Organization).

NASA's UTM Concept of Operations (CONOPS), on the other hand, depends upon each user sharing, through electronic means, the details of the planned flight, so that all users in the airspace will have the same situational awareness of activity in that airspace. This capability is not currently available under the existing air traffic control systems. According to NASA Ames' website, "UTM transforms traditional, human-centric air traffic management into a modern, machine-centric, federated approach allowing distributed management of the airspace wherein disparate entities collaborate to maintain a safe and accessible environment."

NASA led the UTM project along with more than 100 partners across various industries, academia and government agencies, primarily the FAA, all committed to researching and developing this platform.

NASA's UTM research was organized into four phases called TCLs (technical capability levels), each level an increase in complexity and designed with specific technical goals that were intended to demonstrate the capabilities of the system as the research progressed through its phases. Those four TCLs are summarized below (again as quoted from the NASA website).

TCL1: Completed in August 2015 and serving as the starting point of the platform, researchers conducted field tests addressing how drones can be used in agriculture, firefighting, and infrastructure monitoring. The researchers also worked to incorporate different technologies to help with flying the drones safely, such as scheduling and geofencing, which restricts the flight to an assigned area.

TCL2: This was completed in October 2016 and focused on monitoring drones that are flown in sparsely populated areas where an operator cannot actually see the drones they are flying. Researchers tested technologies for on-the-fly adjustment of areas that drones can be flown in and clearing airspace due to search-and-rescue operations or for loss of communications with a small aircraft.

TCL3: Conducted during spring 2018, this level focused on creating and testing technologies that will help keep drones safely spaced out and flying in their designated zones. The technology allows the UAS to detect and avoid other drones over moderately populated areas.

TCL4: From May through August 2019, this was the final level that demonstrated how the UTM system can integrate drones into urban areas. Along with a larger population, city landscapes present their own challenges: more obstacles to avoid, specific weather and wind conditions, reduced lines of sight, reduced ability to communicate by radio, and fewer safe landing locations. TCL4 tested new ways to address these hurdles using the UTM system and technologies onboard the drones and on the ground. These included incorporating more localized weather predictions into flight planning, using cell phone networks to enhance drone traffic communications and relying on cameras, radar, and other ways of "seeing" to ensure drones can maneuver around buildings and land when needed – all while communicating with other drones and users of the UTM system (Figure 4.1).

The NASA UTM research team conceived a ground-breaking way to manage the airspace: a type of air traffic management where multiple parties, from government and the commercial private sector, coordinate their individual capabilities to provide services (Figure 4.2). The UTM research results were conveyed incrementally to the FAA, which continues testing and, with industry partners, has begun implementing the system. By the official end of the project in May of 2021, several efforts had emerged to expand this line of research into other domains, such as managing traffic for the air taxis imagined for urban environments and flights of jets and balloons at very high altitudes (above FL600; Class A airspace) not currently covered by traditional air traffic management services.

This partnership between research and regulation agencies, along with the input of thousands of experts and users, will set the stage for the future of a well-connected sky. Drones will offer many benefits by performing jobs too dangerous, dirty, or dull for humans to do,

Figure 4.1 Drones in flight in downtown Reno, Nevada, during shakedown tests for NASA's Unmanned Aircraft Systems Traffic Management project, or UTM. The final phase of flight tests, known as Technical Capability Level 4, took place from May through August 2019 and studied how the UTM system could integrate drones into urban areas. Credits: NASA/Dominic Hart. A full-colour version of this figure is available as an "Extra" resource by visiting the book's homepage on https://www.nasa.gov/ames/utm.

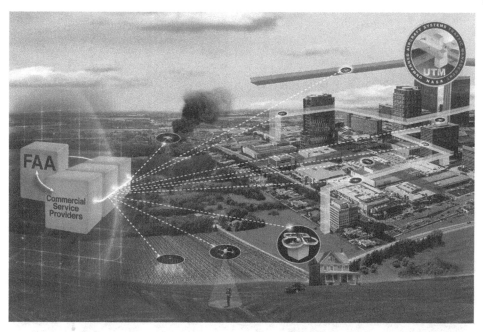

Figure 4.2 UTM system managing various commercial, emergency response, and public safety missions. Credits: NASA. A full-colour version of this figure is available as an "Extra" resource by visiting the book's homepage on https://nari.arc.nasa.gov/utm2021tim.

and NASA is helping navigate towards that future. The NASA/FAA project envisions an airspace management domain that leverages the UTM system to handle all sizes and configurations of UAS in multiple operational scenarios, from small to medium UAS operating at low-level or very low-level airspace, to short-haul deliveries, up to large, high altitude, long endurance UAS operating at or above the commercial flight levels, in Class A and upper Class E airspace (FL600). In addition, the plans call for support of Urban Air Mobility systems operating out of vertiports (Figure 4.3).

The UTM concept is being adopted worldwide with implementations reported in Europe, India, Singapore, Japan, Australia, and other locations. Driven by the work of the UTM team, the International Civil Aviation Organization (ICAO) is issuing guidance to its member states and regulators on UTM framework and principles to ensure global harmonization and interoperability (see Chapter 3).

Independent of, but in partnership with NASA, the FAA initiated its own program to address the demands of the aviation community to develop an unmanned aircraft traffic management model that could be implemented by regulation or statute. This effort was in part stimulated by a series of statutes enacted by the US Congress to deal with the reality that UAS technology was evolving faster than the cumbersome process for promulgating and codifying regulations that had hamstrung the FAA for nearly a decade.

The FAA's NextGen Concept of Operations, V2, "Unmanned Aircraft System (UAS) Traffic Management (UTM)" states:

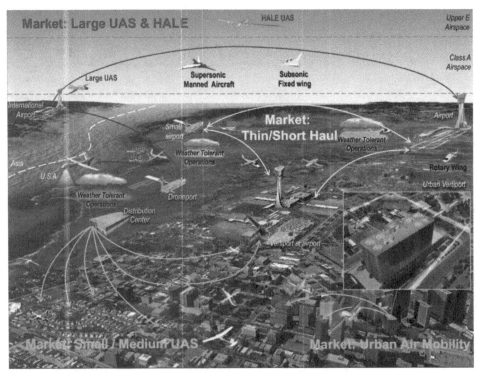

Figure 4.3 Future airspace management domains to leverage UTM System and concepts. Credits: NASA. A full-colour version of this figure is available as an "Extra" resource by visiting the book's homepage on https://www.nasa.gov/aeroresearch/utm-101/.

"This federated set of services enables cooperative management of operations between UAS Operators, facilitated by third-party support providers through networked information exchanges. UTM is designed to support the demand and expectations for a broad spectrum of operations with ever-increasing complexity and risk through an innovative, competitive open market of service suppliers. The services provided are interoperable to allow the UTM ecosystem to scale to meet the needs of the UAS Operator community.

Within the UTM ecosystem, the FAA maintains its regulatory and operational authority for airspace and traffic operations; however, the operations are not managed by ATC. Rather, they are organized, coordinated, and managed by a federated set of actors in a distributed network of highly automated systems via application programming interfaces (APIs). Figure 4.4 depicts a notional UTM architecture that visually identifies, at a high level, the various actors and components, their contextual relationships, as well as high-level functions and information flows. The gray dashed line in Figure 4.4 represents the demarcation between the FAA and industry responsibilities for the infrastructure, services, and entities that interact as part of UTM. As shown, UTM comprises a sophisticated relationship between the FAA, the Operator, and the various entities providing services and/or demonstrating a demand for services within the UTM ecosystem."

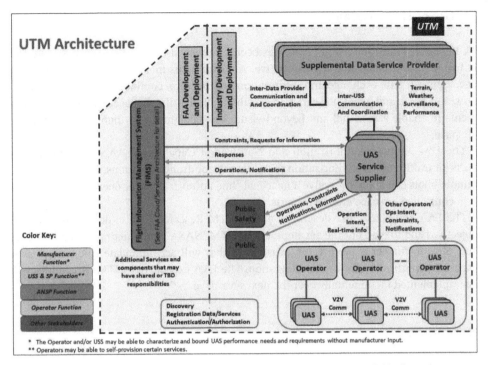

Figure 4.4 FAA Next Gen Notional UTM Architecture. A full-colour version of this figure is available as an "Extra" resource by visiting the book's homepage on https://www.faa.gov/uas/research_development/traffic_management/media/FAA_NASA_UAS_Traffic_Management_Research_Plan.pdf.

The UTM Concept of Operations requires drone operators to inform the UTM system where they want to fly, and the system will autonomously determine if the plan is safe and will not present conflicts with other known or scheduled traffic.

The CONOPs requires a high level of automation and a number of different systems interfacing with one another electronically, with little human intervention. Drone operators may schedule flights using a website or an application on a mobile phone or another device.

This arrangement would be dramatically different from the way the current system works, in that pilots in controlled airspaces are required to remain in contact with air traffic controllers and must get permission for changes in course or altitude.

In replacing or supplementing the existing system, the CONOPs envisions that information about where every aircraft is in the sky at any time would be digitally communicated across the UTM system, creating what Dr. Kopardekar called a "share and care environment."

Companies could be certified by the FAA to provide this network with air traffic management services. The users and participants of the system would be responsible for managing the information flow in the system, taking the burden off air traffic control. Instead of a management by permission philosophy, the UTM network would be managed by exception rather than proscription. Remote pilots will be guided by what they cannot do, such as depart from a geographical zone or geo-fence, as opposed to asking for permission from ATC to fly in a particular block of airspace at an assigned altitude.

Status of UTM Today

"A Research Transition Team (RTT) has been established between the FAA, NASA and industry to coordinate the UTM initiative. Areas of focus include concept and use case development, data exchange and information architecture, communications and navigation, and sense and avoid. Research and testing will identify airspace operations requirements to enable safe visual and beyond visual line-of-sight drone flights in low-altitude airspace.

The Low Altitude Authorization and Notification Capability (LAANC) resource supports air traffic control authorization requirements for drone operations. Through LAANC remote pilots can apply to receive a near real-time authorization for operations under 400 ft in controlled airspace around airports.

The FAA and NASA also developed a joint UTM Research Plan to document research objectives and to map out the development of UTM. NASA is conducting research at UAS Test Sites to further explore UTM capabilities that will accommodate rulemaking as it expands opportunities for drone integration. The FAA expects that UTM capabilities will be implemented incrementally over the next several years."

UTM Vision

"UTM is how airspace will be managed to enable multiple drone operations conducted beyond visual line-of-sight (BVLOS), where air traffic services are not provided.

With UTM, there will be a cooperative interaction between drone operators and the FAA to determine and communicate real-time airspace status. The FAA will provide real-time constraints to the UAS operators, who are responsible for managing their operations safely within these constraints without receiving positive air traffic control services from the FAA. The primary means of communication and coordination between the FAA, drone operators, and other stakeholders is through a distributed network of highly automated systems via application programming interfaces (API), and not between pilots and air traffic controllers via voice."

In response to these several mandates, the FAA initiated the joint project with NASA to create the UTM CONOPS previously discussed, and the UTM Pilot Program (UPP) to define an initial set of industry and FAA capabilities required to support UTM operations.

"UTM services demonstrated in UPP Phase One included: (1) the exchange of flight intent among operators, (2) the generation of notifications to UAS Operators regarding air and ground activities, known as UAS Volume Reservations (UVRs), and (3) the ability to share UVRs with stakeholders, including other UAS Service Suppliers (USS) and the Flight Information Management System (FIMS).

Testing was completed in cooperation with the National Aeronautics and Space Administration (NASA), industry stakeholders, UAS Integration Pilot Program (IPP) participants, and the selected FAA UAS Test Sites."

In Phase Two, the FAA conducted research in the fall of 2020 in cooperation with NASA, industry stakeholders, UAS Integration Pilot Program (IPP) participants, and the selected

FAA UAS Test Sites. UPP Phase 2 moved toward the deployment of remote identification (RID) technologies in increasingly complex environments to enable a UTM ecosystem.

In the next phase, the FAA, NASA, and partners intend to move forward with the primary goal for UPP to develop, demonstrate, and provide enterprise services that will support the implementation of initial UTM operations using a cloud service infrastructure. "These enterprise services will support the sharing of information that promotes cooperative separation and situational awareness."

The UTM Ecosystem is represented in Figure 4.5.

As UAS traffic demand increases in the NAS, it is necessary for the FAA, along with NASA and industry partners, to develop a means to accommodate these operations in a safe and efficient manner. As an initial step in completing UPP Phase 1 – and in support of the UTM ecosystem – the prototype Flight Information Management System (FIMS) has been transitioned to test facilities at the William J. Hughes Technical Center (WJHTC) for integration and testing. FIMS is a central component of the ecosystem providing the FAA with access to UTM data. Figure 4.6 is a graphic representation of a high-level operational concept.

While the components of a UTM system as envisioned by the FAA and NASA differ somewhat from EASA's concept of U-space, the two organizations (and their partners) work together to share information and move towards global harmonization of the UTM/U-space architecture. The differences lie in the regulatory structure needed to accommodate 27 Member States in the European Union, each with its own domestic airspace rules,

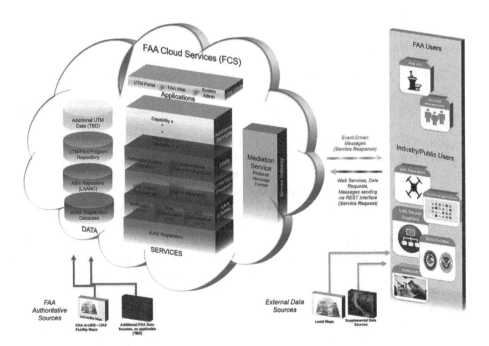

Figure 4.5 UTM Ecosystem (FAA website). A full-colour version of this figure is available as an "Extra" resource by visiting the book's homepage on https://www.faa.gov/uas/research_development/traffic_management/utm_pilot_program/.

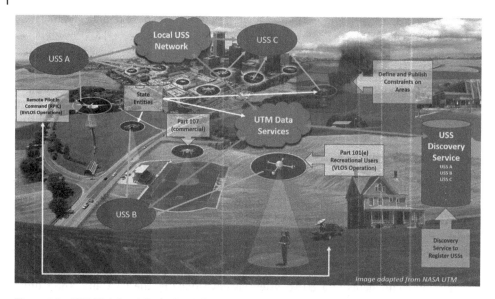

Figure 4.6 UPP High-level Operational Concept, FAA website at UPP Pilot Program. A full-colour version of this figure is available as an "Extra" resource by visiting the book's homepage on https://www.faa.gov/uas/research_development/traffic_management/utm_pilot_program/.

while complying with EUROCONTROL's overarching airspace regulations, whereas in the US there is only one regulatory agency overseeing the vast (and very busy) sovereign national airspace that serves all aspects of American aviation activities.

The core service elements of the notional UTM system (as outlined in the UAS/UTM CONOPS v2.0 document) are:

- **UAS Supplemental Data Service Providers**: Operators and USSs can access Supplemental Data Service Providers (SDSPs) for essential or enhanced services, including terrain and obstacle data, specialized weather data, surveillance, and constraint information. SDSPs may connect to the USS Network or directly to Operators through other means (e.g. public/private internet sites).
- **UAS Service Supplier**: A USS is an entity that assists UAS Operators with meeting UTM operational requirements that enable safe and efficient use of airspace. A USS (1) acts as a communications bridge between federated UTM actors to support Operators' abilities to meet the regulatory and operational requirements for UAS operations, (2) provides the Operator with information about planned operations in and around a volume of airspace so that Operators can ascertain the ability to safely and efficiently conduct the mission, and (3) archives operations data in historical databases for analytics, regulatory, and Operator accountability purposes. In general, these key functions allow for a network of USSs to provide cooperative management of low altitude operations without direct FAA involvement.

    The USS coordinates and distributes to appropriate entities (1) Operator intent, (2) airspace constraint data, (3) weather data, (4) vehicle tracking and conformance data, (5) surveillance data, (6) RID data, and (7) other data critical to safety of flight. This data supports numerous services, including strategic de-confliction, notifications of UVRs, inflight de-confliction/sense and avoid functions, hazard avoidance, and terrain and obstacle clearance, RID, and other value-added services.

To successfully complete these exchanges, USSs must have discovery to FIMS, other USSs, Operators, SDSPs, and public entities (e.g. law enforcement, emergency services, DOD) either directly or via a central inter-USS communication and coordination capability (e.g. the USS Network). Adherence to a common requirement for information exchange within a USS Network (USS-USS) and/or with other specified entities is necessary, along with standard protocols for publishing flight information and other data, and ensures data flow and situational awareness across all participants.

As of September 2020, the FAA has qualified 21 UAS Service Suppliers who are providing their services to over 541 LAANC Enabled Facilities and 732 airports.

Demand/Capacity Balancing: USSs also support collaborative decision-making and conflict avoidance/de-confliction, which promote safety, equitable airspace access, and efficient operations. When users are competing for airspace, USS Operator negotiation capabilities and flight planning tools (e.g. route planning functions, airspace configuration options) are available to support collaborative decision making and/or offer alternate flight intent options that enable equitable airspace configuration solutions designed to optimize airspace equity and access and resolve demand/capacity imbalances.

Exchanges between identified parties require that USSs have discovery to FIMS, other USSs, Operators, SDSPs, and public entities (e.g. law enforcement, emergency services, DOD) either directly or via a central inter-USS communication and coordination capability (e.g. the USS Network). Adherence to an API defined to exchange information within a USS Network (USS-USS) or with other specified entities is also required.

Data Archiving: As the regulator, the FAA monitors Operator compliance with established rules and regulations set forth for the operation; investigates aviation accidents and incidents; collects and analyzes operations data to evaluate whether Operators are meeting agency requirements and goals are being met; and sets the risk for safety and authorizes Operators to operate provided they maintain the established level of safety. USSs will assist the FAA with meeting these responsibilities by archiving requested operation datasets in historical databases for FAA analytics, regulatory, and Operator accountability purposes. USSs must be capable of providing this data upon FAA request.

Remote ID: RID capabilities enhance safety and security by allowing the public, FAA, law enforcement, and other federal security agencies to remotely identify UAS flying in their jurisdiction. Government-qualified RID USSs support RID by receiving and processing network published RID messages, and responding to RID data queries from the general public and authorized officials (e.g. FAA, law enforcement officials).

- **USS Network**: The term "USS Network" is the amalgamation of USSs connected to each other, exchanging information on behalf of subscribed Operators. The USS Network shares operational intent data, airspace constraint information, and other relevant details across the network to ensure shared situational awareness for UTM participants. In the UTM construct, multiple USSs can operate in the same geographical area. The USS Network must implement a shared paradigm, with industry agreed upon methods for de-confliction and/or negotiation, and standards for the efficient and effective transmission of intent and changes to intent. This reduces risk to Operators and improves the overall capacity and efficiency in the shared space. The USS Network is also expected to facilitate the ready availability of data to the FAA and other entities as required to ensure safe operation of the NAS, and any other collective information sharing functions, including security and identification.

- **Flight Information Management System/FIMS:** FIMS is an interface for data exchange between FAA systems and UTM participants. FIMS enables exchange of airspace constraint data between the FAA and the USS Network. The FAA also uses this interface as an access point for information on active UTM operations. FIMS also provides a means for approved FAA stakeholders to query and receive post-hoc/archived data on UTM operations for the purposes of compliance audits and/or incident or accident investigation. FIMS is managed by the FAA and is a part of the UTM ecosystem.

- **National Airspace System/NAS Data Sources:** FAA NAS data sources are connected to the UTM ecosystem through FIMS. This allows for data flow between the UTM community and FAA operational systems, and data access for authorized users, as required. The FIMS interface between the FAA and UTM stakeholders external to the FAA acts as a gateway such that external entities do not have direct access to FAA systems and databases. NAS data sources that FIMS may be connected to for data exchange purposes include sUAS registrations, Airspace Authorizations, operational waivers, and constraints.

Service elements can fall into one of three categories:

1) Services that are required to be used by Operators due to FAA regulation and/or have a direct connection to FAA systems. These services must be qualified by the FAA against a specified set of performance rules.
2) Services that may be used by an Operator to meet all or part of an FAA regulation. These services must meet an acceptable means of compliance and may be individually qualified by the FAA.
3) Services that provide value-added assistance to an Operator, but are not used for regulatory compliance. These services may meet an industry standard, but will not be qualified by the FAA.

The following is a list of UTM services that are anticipated to be provided by this network of USSs. The list is not exhaustive and may be added to as the technology evolves.

- Remote ID (services to identify UAS operating in the NAS)
- FAA Messaging (exchange capabilities for regulatory/policy requirements)
- USS Network Discovery (discover relevant active USS providers and operations in a specific geographic area)
- Operator Registration (for vehicle owners to register data)
- Airspace Authorization (from airspace authority/ANSP to UAS operator)

- Constraint Management (operational constraints from public safety activities)
- Operator Messaging (position reports, intent and status information)
- Strategic Deconfliction (arranges, negotiates, and prioritizes intended ops volumes)
- Conformance Monitoring (real-time alerting of non-conformance)
- Conflict Advisory and Alert (real-time monitoring and alerting)
- Dynamic Reroute (real-time modifications to intended operation volumes/ trajectories)
- Operation Planning (all aspects of flight planning)
- Flight Dispatch (commercial ops under part 135 or special authorization/certification)
- Weather (forecast and real-time)
- Mapping (airspace constraints, restrictions, NOTAMS, and ground constraints)
- Communication/C2 (RF command and C2 capabilities to operators)
- Surveillance (supporting air risk assessment, flight planning, and flight operations)
- Navigation (surveying, coverage maps, real-time availability, integrity, quality of service)
- UAS System Monitoring (health and status of the UAS system components)

Participation

All UAS Operators not receiving ATC separation services are required to participate in UTM at some level using applicable services to meet the performance requirements of their operations.

Performance Authorizations

Operations in UTM will need to rely on the established construct in manned aviation (instrument approach procedures, air traffic management discretion, and RNAV/RNP routes, for example) through the issuance of Performance Authorizations. UAS operations are expected to vary greatly in CNS (communications, navigation, surveillance) performance considering the many aircraft types and their intended operations. The CONOPS expects that the variance in performance will be managed by the USSs when providing different services. The USSs must account for the variance while maintaining safety and equity in the airspace.

It is not expected that the regulator will dictate specific CNS performance for airspace that is not under ATC control. CNS performance requirements will either be a product of the operator-specific safety case or determined more nimbly (real-time?) by the USSs to effectively perform the offered services.

In the UTM ecosystem, the regulator has the role of ensuring interoperability of the system participants. Interoperability in UTM focuses on how data is exchanged and interpreted by those involved. A common understanding of CNS requirements between actors is critical to the overall safety case. Depending on the overall risk of the underlying operation, substantiating data may be required of the applicant before operations can be authorized. Authorizations must be secured before conducting a class or type of operation. In the US, such applications must be made to the FAA.

Airspace Authorization

Airspace authorization must also be obtained from the FAA. A Performance Authorization substantiates an Operator's ability to meet flight performance capabilities in their intended area of operation, while an Airspace Authorization grants access to operate in controlled airspace and provides the air traffic facility with jurisdiction over the airspace access to information about operations being conducted.

The other items on the list of services should be self-explanatory. The list is essentially consistent with its equivalent in the European U-space environment, with the major difference being the role of the regulatory authorities (numerous CAAs in Europe, one in the US).

Recent Developments in UAM/AAM

An article published on 7 October 2021 on the Unmanned Airspace news forum reported:

"The US Federal Aviation Administration (FAA) will deliver an updated version of its Concept of Operations (ConOps) for Urban Air Mobility (UAM) in the first quarter of 2022," said Steve Bradford, FAA Chief Scientist for Architecture and NextGen Development, speaking at the AUVSI webinar *Reimagining Mobility: Update on UAM & AAM on 6 October 2021*. "We have foundational work still to do and we will be engaging with industry on this next refinement." The work includes workshops to "flesh out" concepts in partnership with the National Aeronautics and Space Administration (NASA) Aeronautics Research Institute (*NARI*).

The FAA released UAM ConOps Version 1.0, developed in association with NASA and industry, in July 2020, to describe early stage, low density urban air taxi operations in the US National Airspace System. "UAM ConOps 1.0 brought into existence UAM corridors. While a good starting point, the requirement to fully participate in all airspace classes becomes restrictive."

NASA AAM Mission Manager David Hackenberg said: "The FAA document drives what we want to do at NASA from an R&D perspective. It lays the groundwork that we can build against and sets the pathway for industry development."

Describing the key differences between UAM ConOps 1.0 to 2.0, Bradford said the updated version will be fully coordinated. "We believe we can reduce the impact on non-UAM participants, and we will probably have fewer corridors." The corridors will vary according to the class of airspace with participation requirements for crossing traffic depending on the airspace class. Additionally, there will be alternative mitigation strategies for crossing traffic.

The FAA is aiming to introduce greater flexibility to the corridor designs, for example by allocating greater tactical separation responsibilities to the ecosystems. "Safe operations occur based on a combination of strategic deconfliction and tactical separation. In UAM, strategic deconfliction is part of the ecosystem of Providers of Services for UAM (PSU)." However, this is likely to introduce large buffers, so tactical separation will also be performed by the UAM aircraft with support from the PSU network. "This is an area where we are need further research," he added.

The FAA identifies a number of focus areas to refine areas of responsibility and performance requirements. Among these, defining UAM operational intent is important to manage traffic flow within the corridors, along with further analysis of different traffic mixes. "4D trajectory is probably not enough, we need to establish what will be required." Vertiports will service a range of aircraft – flying in corridors as well as controlled airspace – and will not operate in a sterilized environment. The design of the corridors is also significant to ensure they interface seamlessly with Unmanned Traffic Management (UTM) and Air Traffic Control (ATC).

"We looked at the impact on airspace management operations and there was always some impact, for example for general aviation and helicopters. There is no such thing as one flat superhighway in our analysis." In case studies conducted by FAA in the Dallas Fort Worth area, it proved difficult to create straight lines between locations due to other activities ongoing in the airspace. "We tried to explore concepts that minimize the impact on non-UAM crossings while retaining flexibility and access for UAM to provide scalable solutions and optimize all integration objectives." As a consequence, while corridors are expected to be needed in all classes of airspace, the performance requirements for UAM and non-UAM aircraft are likely to vary. For example, in Class B airspace non-UAM aircraft will need to provide operational intent in order to participate, while in less complex Class C airspace ADS-B will be a minimum requirement.

Another significant factor is cybersecurity, where "so much information exchange is with industry, and stays with the industry," added Steve Bradford.

Conclusion

The next chapter explores UAS airspace integration efforts in other parts of the world not subject to European Union or US regulations. That includes BVLOS UAS and VLOS UAS operations, as well as manned aircraft.

References

Barnhart, R., Marshall, D., and Shappee, E. (editors/authors) (2021). *Introduction to Unmanned Aircraft Systems*, 3e. Boca Raton: CRC Press.

FAA. website available at: https://www.faa.gov/uas/research_development/traffic_management/

FAA NextGen Concept of Operations v2.0 Unmanned Aircraft System (UAS) Traffic Management (UTM), available at: https://www.faa.gov/uas/research_development/traffic_management/media/UTM_ConOps_v2.pdf

The FAA Modernization and Reform Act of 2012 (hereafter FMRA), Pub. L. 112-95, Title III – Safety, Subtitle B – Unmanned Aircraft Systems.

The FAA Extension, Safety, and Security Act of 2016 (Pub. L. 114-190) Title II, Subtitle B-UAS Safety.

The FAA Reauthorization Act of 2018 Title III, Subtitle B Unmanned Aircraft Systems (Pub. L. 115–254) Sections 351, 376 and 377.

Title 49 United States Code § 40103.

UAS Traffic Management (UTM) Research Transition Team Plan (RTT) Plan. (2017). Available at: https://www.faa.gov/uas/research_development/traffic_management/media/FAA_NASA_UASTraffic_Management_Research_Plan.pdf

Unmanned Aerospace *FAA "on track to issue UAM concept of operations 2.0 in first quarter of 2022" – Chief Scientist Steve Bradford* by Jenny Beechener. Available at: https://www.unmannedairspace.info/emerging-regulations/us-initial-urban-air-mobilityconops-modeled-on-existing-utm-operations (accessed October 11, 2021).

Unmanned Airspace News. *US releases initial Urban Air Mobility ConOps, inspired by UTM operations*, July 1, 2020 available at: https://www.unmannedairspace.info/emergingregulations/us-initial-urban-air-mobility-conops-modeled-on-existing-utm-operations (accessed October 2, 2021).

5

Global Airspace Integration Activities

The European Union and the United States are not the only regions taking steps to integrate unmanned aircraft operations into their respective national airspaces. The two major efforts have been in Europe and the United States, but other nations have participated or observed in some fashion the progress that has been made in those geographical areas. This chapter will examine, at a top level, how other countries have reacted to the growth of remotely piloted aircraft in their jurisdictions. The countries selected are just a sampling of airspace integration activities around the world.

Australia

The Civil Aviation Safety Authority (CASA) creates regulations that cover safety, operating rules, drone accreditation, remote pilot licensing, exclusions for very small UAS and flying over one's own property, operations over open-air assemblies of people, drone registration and markings, RPA operator's certificates, and certified training providers, among others. CASA adheres to Australian government requirements for rulemaking in an eight-step process: Initiation and Planning; Initial Consultation; Formal Consultation; Legal Drafting; Application of Regulatory Best Practices; Legislative Approval; Implementation; and Project Closeout and Review. Public engagement, collaboration, and consultation are required at all stages. Aviation Safety Rules are contained in the Civil Aviation Act of 1988, Civil Aviation Regulations 1988 (CAR), and Civil Aviation Safety Regulations 1998 (CASR). All aviation regulations may be accessed on the Australian Government Federal Register of Legislation.

The new *Transport Safety Investigation Regulations 2021* will repeal and replace the *Transport Safety Investigation Regulations 2003*. These new regulations will continue to require the reporting of certain transport safety occurrences to the ATSB as immediately or routine reportable matters. The main changes to these regulations will be the introduction of updated requirements for operators of certain types of remotely piloted aircraft (RPA) to make reports to the ATSB after a mishap or incident.

Recognizing the range of different types of RPA and their uses, the regulations will categorize RPA as Type 1 or Type 2 RPA. RPAs that have been certified under relevant airworthiness standards (type certification) as large (greater than 150 kg) and medium RPAs

UAS Integration into Civil Airspace: Policy, Regulations and Strategy, First Edition. Douglas M. Marshall.
© 2022 John Wiley & Sons Ltd. Published 2022 by John Wiley & Sons Ltd.

(more than 25 kg but not more than 150 kg) are defined as Type 1 and are an emerging form of commercial aviation that will benefit from investigation into systemic safety issues to help prevent future accidents.

In contrast, Type 2 RPAs are defined as those RPA that are not Type 1, excluded or micro-RPA (gross weight less than 250 grams), and will have fewer reporting requirements. This distinction is made on the basis that ATSB investigations are unlikely for these operations unless there is serious risk of harm to people or a significant third-party property.

Australia has aggressively pursued opportunities in Advanced Air Mobility and Urban Air Mobility concepts. CASA will be releasing a draft RPAS and AAM Regulatory Paper at the end of 2021 for industry input (Spence 2021).

The Federal Department of Industry, Transport, Regional Development and Communications (DITRDC) will be proceeding with the release of a call for proposals in late 2021 for PARTNERSHIPS to undertake advanced air mobility demonstrations in Australia during 2022– 2023. Primary focus of the PARTNERSHIPS program will be on AAM operations in regional Australia and AAM global supply chain roles for Australian industry, including local manufacturing opportunities.

The AAM infrastructure plan framework is in development to provide guidance to planners at the national level and to clarify processes for industry as to how they need to go about getting approvals. The Federal Government wants a seamless, single national process and framework across the country, addressing federal, state, and local needs. Different aircraft types are likely to increase complexity of all dimensions of safety, noise, and security requirements. Airspace design and regulation will be an ongoing challenge (e.g. the interaction of RPAS and eVTOLs carrying people). Autonomous operations will be enabled by a national UTM system, which is currently under development. FIMS (flight information management system) is the foundation for the UTM system, and that work is in development as well.

The Aviation Safety Advisory Panel incorporates a Technical Working Group (TWG) to co-design AAM elements. TWG has four key questions to respond to: (1) where should CASA regulate, (2) where can tech be leveraged to assist with regulation, (3) where are non-tech solutions best suited, and (4) how will the Australian system link and harmonize with the rest of the global AAM system?

CASA recognizes it needs to think differently as a regulator. The historical model of an aviation regulator is prescriptive, which reflects the early years of manned aviation. Emergence of concepts such as crew resource management and Safety Management Systems started to change that model. The contemporary standard for a regulator is one who will learn and continually evolve. Equivalent safety is a concept that is critical to providing that freedom to work within the safety rules.

CASA will work with current and prospective operators, the latter being essential as a lot of RPAS and AAM developers and operators are from outside traditional aviation circles. More than 2200 organizations now hold a ReOC (Remote Piloted Aircraft Operator's Certificate), ranging from sole operators to multinational operations, while 30 000 RPAS are registered in Australia, 22 000 remote pilot licenses have been issued, and the number is growing at a rate of 300 per month. The growth rate of the industry stimulates the need for CASA to ensure streamlining of approvals while ensuring appropriate safety standards. Digital airspace authorizations near airports are being tested at present (Canberra/Perth/Brisbane) by automatically assessing applications and making a real-time determination via a participating CASA-verified drone safety application. The automated process reduces approval times from weeks to minutes, allowing commercial operators to secure more business faster and at a reduced cost.

Digitization is clearly critical to the future of authorizations, as is standardization of application and approvals processes. This is entirely consistent with both the NASA/FAA UTM process and the U-space concept in Europe.

Historically, each application for approval is considered on a case-by-case basis using the SORA model. This has shifted to standard scenario-based models, where the applicant tailors its proposed operation to that standard scenario, leaving only a need to review those elements that differ, rather than the entire application. Continued evolution of regulations will include all elements of AAM – aircraft, infrastructure, autonomy, safety, noise, etc. The historical approach to regulation has been "crawl, walk, run." However, CASA does not intend to wait for the full process of developing regulations to allow an AAM industry to come into being.

The Australian Association for Unmanned Systems (AAUS) created a UTM Working Group made up of nine industry representatives. The tasks and highlights of this organization's activities are:

- Promoting and facilitating development of Australian UTM solutions in concert with government and international opportunity for Australian UTM industry.
- Fostering UTM knowledge development across the Australian ATM and UAS/RPAS sector.
- In August of 2020, Airservices Australia released a Request for Information (RFI) for the development of a Flight Information Management System (FIMS) Prototype relating to the development and evaluation of a flight information management system for UAS Traffic Management.
- AAUS works to advance UTM in Australia, which they consider to be a crucial enabler for large scale integrated operations, particularly for UAS and advanced air mobility (AAM) operating around their cities.
- AAUS believes it is essential that policy and a regulatory framework for UTM be developed before discussion on specific implementations or technical solutions. Work to develop the necessary regulatory framework for UTM should commence as a matter of urgency. Industry stands ready to support CASA in this undertaking.
- AAUS strongly believes that more proactive and transparent stakeholder engagement is essential and that this engagement should be inclusive of all airspace users and not just the #drone/#RPAS sector.
- Industry welcomes the chance to share in and evaluate UTM innovations with Airservices Australia towards informing this much needed discussion on the safe and efficient integration of UAS and AAM into Australian airspace (La Franchi 2021).

Brazil

Brazil's CAA (Agéncia Nacional de Aviação Civil) implemented Brazilian Civil Aviation Regulation No. 94/2017 Special (RBAC-E No. 94/2017) in 2017 as a supplement to drone operating standards established by the Department of Airspace Control (DECEA) and the National Telecommunications Agency (ANATEL). The rules create two categories of

UAS: model airplanes and RPA. Remote pilots of Class 1 or 2 RPA, or anyone intending to fly above 400 ft AGL, must have a valid license and license issued by the ANAC. ANAC also issues Special UAS-CAER Airworthiness Certificates. ANAC oversees pilot licensing and qualifications, project authorizations, registration of aircraft and experimental airworthiness certificates, and design authorizations. The essential rule for model aircraft and RPAs is to keep the required distance from third parties (30 m), only operate one aircraft at a time, and follow the rules established by DECEA and ANATEL. The Brazilian Aeronautical Code (CBA) Law No. 7565 is the overarching aviation legislation that controls all the subcategories, including drones. Brazil recognizes the authority of ICAO Doc. 10019 Manual on RPAS (discussed in Chapter 4).

Brazil is blazing a trail in drone use, using unmanned vehicles in agriculture, food delivery, search and rescue, and inspection. There are 80,000 drones flying legally today, of which 35,000 conduct commercial operations. Visual line of sight (VLOS) flight authorizations below 400 ft are accessed free, in real-time, using the regulator's SISANT registration portal. The government is designing a framework to enable beyond visual line of sight (BVLOS) operations at scale as part of a wider Unmanned Traffic Management (UTM) concept.

BVLOS operations are authorized on a case-by-case basis, with approximately 60 unmanned aircraft approved for regular BVLOS operations. Applications to fly BVLOS are made using DECEA's SARPAS platform, which sets strict limits around airfields, critical infrastructure, and over people. "Under ANAC's regulatory agenda, the first work stream enabled VLOS from 2018 and BVLOS by authorization from 2020. A second work stream during 2021/22 is focused on simplifying drone registration and a third work stream is developing technical criteria for the new operational environment to support UTM," according to Ailton José de Oliveira Jr, ANAC's Coordinator Drones and New Technologies Group.

SARPAS has received 500,000 access requests since the concept was introduced in 2016. "The next step is phased development of the SARPAS front end. The main goal is the safety of all airspace users." (Jorge Alexandre de Almeida Regis, DECEA Head of military Op Subdivision, said that DECEA proposes a set of four levels of service to encompass controlled and uncontrolled airspace.) "Tests will include simulated flight tests with participation from all stakeholders to demonstrate interoperability with ATM."

While the Brazilian landscape is uniquely suited to drone use, it also faces challenges. It is the fifth country worldwide in terms of helicopter usage, sharing the same low-level airspace as drones, and internet access is limited.

Brazil was represented at ICAO's Drone Enable 2021 Symposium. The BRAZILIAN FOCUS panel highlighted the entire process related to UAS operations in Brazil and involved all the key stakeholders, including operators and Brazilian Authorities. Attendees were provided with an understanding of the main goals and the challenges during the various processes. In addition, the panelists discussed the necessity of implementing a viable UAS Traffic Management system as a key enabler for UAS scalability.

Canada

Canadian aviation regulations (CARs-SOR/96–433) are managed by Transport Canada. The Canadian rules and rulemaking processes are similar to US regulations, but some were promulgated and implemented years before the FAA put a set of rules in place in the US.

Remotely Piloted Aircraft Regulations, which are found in Part IX (Sections 900.01–903.01) and Standards (Sections 921–922). An SFOC (Special Flight Operations Certificate) may be required depending upon whether the flight is recreational or commercial, the drone's weight, and if particular exemptions may apply. Recreational UAS under 35 kg in weight do not require a certificate, whereas all aircraft heavier than 35 kg require a certificate regardless of use. sUAS (between 250 g and 25 kg) may be exempted if the operator is able to follow strict safety conditions. The classification system is based upon risk, and closely resembles US drone regulations.

In Canada, UTM is also referred to as "RTM," Remote Piloted Aircraft Systems (RPAS) Traffic Management. By any name it is a complex system of technologies working together,

In late 2020, Canada's RTM initiative, a joint program run by Transport Canada and Nav Canada, selected two teams to carry out official service trials to test unmanned traffic management operations.

Starting in a 425 km corridor in Northern Alberta, the partners will be conducting Cellular-enabled UTM and Sensing Data Services in conjunction with one of North America's leading energy infrastructure companies. The trial aims to generate data and experience that will be used by Transport Canada in the development of regulations for BVLOS (Beyond Visual Line Of Sight) drone operations and the delivery of commercial cellular-enabled UTM services. Early tests of the system architecture under VLOS (Visual Line Of Sight) regulations have been conducted.

The telecommunications network will gather data on cooperative air traffic, enabling the team to benefit from comprehensive air traffic awareness (including at low altitudes).

NAV CANADA is Canada's air navigation service provider. For advanced operations only, to operate in controlled airspace (Classes C, D, or E) operators need to ask NAV CANADA for an RPAS Flight Authorization. In anticipation of continued growth in drone traffic in Canada, NAV CANADA announced plans to launch NAV Drone in the spring of 2021. This is NAV CANADA's official instrument for drone flight authorization applications, available on desktop and mobile devices. NAV Drone will help provide pilots and drone operators with the tools they need to safely fly their drones in Canadian airspace. NAV CANADA selected Unifly as the Unmanned Traffic Management (UTM) technology provider in February 2020.

Currently, all applications for authorization to fly drones in controlled airspace are manually approved by NAV CANADA and, depending on their complexity, approvals can take up to 48 hours to process. With the NAV Drone, NAV Canada predicts that approximately 70% of requests will be approved automatically in the app, reducing wait times for pilots and drone operators. NAV CANADA employees will therefore be able to focus on the growing number of requests for access to controlled airspace that require detailed reviews.

Nav Drone Pilots and operators will be able to plan and schedule drone flights, receive clearance responses from NAV CANADA, find out where they can and cannot fly, and obtain important information regarding Canadian airspace.

China

China's national civil aviation authority is the Civil Aviation Administration of China (CAAC). All permissions to operate unmanned aircraft in China are granted by the CAAC. Licensing is required for commercial operations. Any drone weighing more than 250 g must be registered with the CAAC. Drones weighing more than 116 kg require a pilot's license and UAV certification. Operational restrictions are similar to other nations (no BVLOS operations, no flights above 120 m AGL, no flights in densely populated areas, no flights in airport environments, military installations or other sensitive locations, all flights are prohibited from operating in designated no-fly zones, and no flights may be conducted in controlled areas without CAAC approval). All drone operators are required to carry third-party liability insurance.

The classification of UAS and the supervision measures set forth in the UAS Operation Rules are consistent with the direction of the development of the rules of the Federal Aviation Administration in the US (FAA) and the European Aviation Safety Agency (EASA).

A key component of China's UAS Operation Rules is an online, real-time supervision (or surveillance) system, including the "UAS Cloud" and the "electronic fence" (geo-fence). The "UAS Cloud" is a dynamic database management system that monitors flight data (including location, altitude, and airspeed) in real time and has an alarm function for UAS flying into the electronic fence. The electronic fence is a software and hardware system in which specific geographic zones are identified as prohibited, and then function to stop aircraft from entering such areas.

For UAS of Types III, IV, VI, and VII, the installation and use of the electronic fence and connection to the UAS Cloud are required, and operators must report every second when in densely populated areas and every 30 seconds when in non-densely populated areas. Flight data recorded in the system should be stored for at least three months.

For UAS of Types II and V, only those operated within key areas and airport clear zones are required to install and use the electronic fence, connect with the UAS Cloud, and report every minute.

For UAS that are not required to connect with the UAS Cloud, the UAS Operation Rules still require clear information to be carried on each UAS, such as the name and contact details of the owners, for tracking and identification purposes. This set of requirements is similar to the registration requirements imposed by the US FAA in its Final Rule for light UAS weighing from 0.25 kg to 25 kg, and shows that the CAAC authority is, in this respect, following the global trend to improve the administration of UAS in this size category.

Regardless of whether a UAS Cloud system is used, a UAS pilot-in-command is still responsible for ensuring that the UAS is operating in accordance with the requirements of relevant authorities and preventing the aircraft from entering into any restricted areas.

Restricted areas under the UAS Operation Rules include airport obstacle control surfaces, prohibited areas, unauthorized restricted areas, and danger zones. Which specific areas qualify as restricted in practice are not set in advance, but are instead within the charge of the relevant authorities to determine. If a UAS is connected to the UAS Cloud, the restrictions will be shown in the UAS Cloud. If a UAS is not connected to the UAS Cloud, then its user should consult with the relevant authorities about the restricted areas.

In late 2020, the Chinese government announced its intention to accelerate the country's adoption of urban air mobility (UAM) operations. In a 30 November 2020 statement, the General Office of the State Council declared that it is bringing UAM development into China's National Strategies "to formulate relevant policies and standards to promote the healthy development of the industry."

China anticipates that these policies and standards will provide the regulatory foundation that "should pave the way for China to become the world's largest UAM market." The circular called for advancing the legislative process to establish regulations covering the operation of unmanned aerial vehicles for early applications such as fire-fighting and various industrial activities.

One China-based eVTOL aircraft developer interpreted the announcement as a tacit endorsement of its plans to expand operations with its two-seat 216 Autonomous Aerial Vehicle. Claiming to be the only company currently authorized by the Civil Aviation Administration of China (CAAC) to seek airworthiness approval for new aircraft intended for UAM applications, the company indicated that the new policy will help it to progress beyond the trial operations now being conducted in several locations. (It should be noted that this company, and others in the same market, were profiled in a public television "NOVA" production aired in the US on 26 May 2021, in a segment entitled "Great Electric Airplane Race.")

The State Council's circular does not include a specific timeline for introducing complete requirements for airworthiness approval, type certification, and operational regulations. It also provides no indication as to when or how companies other than the one just mentioned will be permitted to advance competing plans for UAM service.

Earlier in 2020 the CAAC announced China's first Unmanned Civil Aviation Zones, which reportedly provide some structure for initial eVTOL aircraft flight trials. The accompanying guidance issued proposed so-called "convenient channels" to support arrangements for "airworthiness, operation, air traffic control, and business licensing."

These UAM efforts will require some form of a UTM architecture similar to that of the US and Europe to safely manage airspace integration, particularly low-level and very-low-level UAS mixing with remotely piloted or piloted eVTOL air taxis with human occupants on board.

France

The French Transport Ministry and the Ministry of the Ecological and Inclusive Transition control drone policy and practice in France. AlphaTango® is the new name for the portal set up by the DSAC for users of piloted aircraft to register their aircraft, obtain training certificates, and secure permission to operate commercial UAS. France follows European Union regulatory guidance. The Ministry publishes a guide (in French) for all users of unmanned aircraft in France (Aéronefs Circulant Sans Personne a Bord: Activitiés Particulières). Generally, the same basic operational rules apply as found elsewhere: no BVLOS flights, operate under 150 m AGL, daytime flights only, no urban flights, no flights over people, no flights in restricted areas, respect privacy of others, no commercial use of images without permission, mandatory liability insurance, aircraft registration, remote pilot certification (private pilot, or ultralight or glider license), and no flights in airport environments. French law describes seven different categories of aircraft, based upon mass, and four different operational scenarios. BVLOS flights and flights over congested areas may be permitted for aircraft weighing less than 4 kg. Commercial operations require a licensed civil pilot and permission to operate. All UAS weighing more than 800 grams must be registered and must carry audible warning devices and navigation lighting. Penalties for violations of the regulations may include imprisonment for up to one year and a €90 000 fine.

France is a part of the European Union and is an EASA Member State. As such, it is presumed that it will conform its RPAS regulations and standards to the evolving EASA and EUROCONTROL plans for U-space implementation. Cross-border authorizations and harmonization of operating and certification rules depend upon cooperation and consistency among the Member States.

In line with that policy, in September 2021 the French transport ministry, in partnership with the Defense Innovation Agency (AID), issued a call for projects to accelerate the implementation of the "U space" program first introduced by EASA in 2018.

The call is the fourth part of the Transport Innovation Agency (AIT) "Propulse programme" announced on 30 August 2021 and the first to also be supported by the Defense Innovation Agency (AID). It is a response to both civil and military issues and aims to create a coherent and effective airspace environment.

The "U-space" program designates the future management of air traffic for drones, with a high level of automation and digitization, guaranteeing a safe integration of drones in airspace from a safety and security point of view, but also respectful of the environment and protection of privacy.

The French "U-space Together" call for projects aims to experiment on a very large scale with solutions for air traffic management services for drones, at very low altitude, in almost the entire territory of mainland France. It has five packages aimed at meeting a variety of technical and operational challenges linked to the rise of drones in France and in Europe.

Germany

Germany's UAS regulations went into effect on 7 April 2017. The Federal Aviation Office (LBA) is the national civil aviation authority responsible for the issuance of certificates of proof of sufficient knowledge and skills for the operators of unmanned aerial vehicles. Inquiries on details of the requirements may be sent to AST@lba.de. The regulations are not currently available in English on the official LBA website. Highlights include UAS weighing less than 5 kg may be flown without a permit, but those weighing more than 2 kg require a licensed remote pilot. UAS may only be flown above 100 m AGL with LBA permission and UAS over 5 kg require a permit to fly at night. UAS weighing more than 250 g must be registered and marked with a fireproof data plate. A certificate of knowledge is required to operate a UAS heavier than 2 kg and UAS may not be flown over crowds, industrial areas, disaster areas, and other designated sensitive locations. Liability insurance is required. It should be noted that Germany is a federal republic with 16 states, and there is nothing in the regulations that prevents any of those states from enacting their own drone regulations so long as they do not conflict with LBA, EASA, and EUROCONTROL rules.

Germany is a Member State of the European Union and will presumably adhere to the European Plan for Aviation Safety and other EU implementing regulations with regard to U-space and RPAS airspace integration. For reference there is an EASA document posted on the EASA website entitled "Cooperative Arrangement Between the European Aviation

Safety Agency and the Luftfahrtamt Der Bundeswehr Concerning Aviation Safety." The Agreement contains the following language:

Section 4 Core Areas of Cooperation

The cooperation between the Participants will comprise the Airworthiness, Flight Operations, Licensing/Organization Approvals/Recognition and Aviation Medicine domains and inter alia will focus on the following:

    a) Aviation safety;
    b) Incident reporting systems;
    c) Cybersecurity in aviation;
    d) Remotely Piloted Air Systems (RPAS), including airworthiness and integration into the airspace;
    e) Performance Equivalence, and
    f) Air Traffic Management (ATM) with a particular focus on Communication, Navigation, and Surveillance (CNS).

Ireland

The Irish Aviation Authority (IAA) promulgated drone regulations in "Small Unmanned Aircraft (Drones) and Rockets Order S.I. 563 of 2015." The regulations were updated in 2021. All drones continue to be regulated under national legislation pending implementation of the new European Union rules, but operators are encouraged to make themselves familiar with EU Regulation 2019/947. Ireland's national drone regulations mirror those of most other European countries: avoid hazardous operations, no flights over assemblies of people (12 or more), no flights beyond 300 m from the operator, no flights above 400 ft AGL, no flights in urban areas, no flights near airports or restricted or sensitive areas, and no flights in controlled airspace. Specific Operating Permission (SOP) is required to deviate from any of the limitations, and the operator must receive training from an approved training facility and a Pilot Competency Certificate is required. All drones over 1 kg in maximum takeoff weight (MTOW) and operated above 15 m AGL must be registered. No permission is required to operate a drone commercially. Model aircraft are treated the same way as commercial drones.

    Drone operations are to be conducted according to the Commission Delegated Regulation (EU) 2019/945 and Commission Implementing Regulation (EU) 2019/947 (as amended). The Irish Aviation Authority (IAA) supervises and implements the Regulation in Ireland. The 2021 revision to the IAA regulations were intended to create a harmonized drone market in Europe, while preserving the highest level of safety. In practice, this means that once a drone operator has received authorization from its state of registry, they will be allowed to freely circulate throughout the European Union. According to the level of risk involved, this new legal framework introduces the three categories of drone operations (Open, Specific, and Certified), discussed at length in Chapter 2. Ireland will have the same criteria for establishing U-space and AAM/UAM operations as the other Member States in the EU.

Italy

Italy has implemented its own domestic regulations relating to drones. The Italian Civil Aviation Authority (ENAC) states that it is following the lead of EASA and EUROCONTROL in working on operations in critical areas (ATZ) and operations BVLOS in non-segregated airspace. Procedures for drone activities are included in ATM-5 circular of 23 July 2013, which is only available in the Italian language.

Italy's RPAS regulations are contained in "MEZZI AEREI A PILOTAGGIO REMOTO" (Remotely Piloted Aerial Vehicles), revised in 2018. The regulations are set forth in eight sections: Section I – General; Section II – Remotely Piloted Aircraft System with aircraft having operating take-off mass of less than 25 kg; Section III – Remotely Piloted Aircraft System with aircraft having operating take-off mass of more than or equal to 25 kg; Section IV – Provisions for piloting RPAS; Section V – Rules of circulation and use of airspace; Section VI – General Provisions for RPAS; Section VII – Model Aircraft; and Section VIII – Final provisions. The ENAC website provides for an English translation from Italian.

From the ENAC website:

> "According to Regulation 2019/947, all certificates relating to UAS remote pilots, issued after 31 December 2020 by one of the national authorities subject to EASA regulation, are automatically recognized as valid by all other national authorities. For this reason, a remote pilot in possession of an EASA certificate can operate freely within each EASA member state, in compliance with European legislation and in compliance with any additional national restrictions, referring to a section of this website entitled 'where can I fly my drone?').
>
> In view of the current temporary absence of a single European database, individual civil aviation authorities have created their own internal databases for the registration of certificates issued at national level. These databases are not interoperable and do not allow a rapid verification of foreign certificates by ENAC."

In order to allow the control of the prerequisites for obtaining a higher level certificate to a citizen in possession of a remote pilot certificate issued by another national authority (e.g. a remote pilot in possession of a foreign A1/A3 certificate who wants to achieve an A2 in Italy), foreign certificates must be forwarded to ENAC through the "serviziweb" portal, for the necessary checks and for their acquisition in the ENAC database.

In summary, Italy has attempted by bring its domestic drone regulations in compliance with EU regulation and policy. There are no specific references to U-space or airspace integration in the current regulations, but the referral to Regulation 2019/947 and its amendments incorporates the EU efforts to implement U-space. It remains to be seen what the Italians may do with AAM/UAM, particularly in their more popular tourist cities such as Venice, Florence, Milan, and Rome, where there are ongoing efforts to limit the number of tourists crowding their streets (especially under the threats posed by the COVID-19 pandemic).

Japan

Article 87 of Japan's Civil Aeronautics Act is a two-paragraph treatment of pilotless aircraft, and merely states that the Minister of Land, Infrastructure, Transport and Tourism (MLITT) must grant permission for UAS flights, and may impose flying restrictions on the aircraft when "he/she deems it necessary to prevent dangerous effects on other aircraft." An amendment to the Aeronautical Act setting new rules for unmanned aircraft went into effect on 10 December 2015. The new rules define the airspace into which drone pilots must seek permission to enter from the MLITT (around airports, airspace above 150 AGL, and above densely inhabited districts (DIDs)). As of 18 September 2019, new rules prohibit operating under the influence of alcohol or drugs, require a preflight "action" (presumably checklist and system validation), operating so as to prevent collisions with airplanes and other UAs, and operating in a careless and reckless manner. Night-time operations, BVLOS operations, operating within 30 m of persons or objects on the ground, operating over event sites or gatherings of people, transporting hazardous materials, and dropping objects require approval from the Regional Civil Aviation Bureau in advance. Permissions must be sought 10 days in advance of operations from the MLITT. Violators may be liable for a fine of up to 500 000 yen (about $4591 USD), and up to one year of imprisonment if operating under the influence of alcohol or drugs.

- RPAS may not be flown in the following ways without special permission from the Minister of Land, Infrastructure, Transport and Tourism: 150 meters (492 ft) above ground level; near airports; and above densely populated areas, as defined by the Ministry of Internal Affairs and Communications.
- To request special permission, operators must submit an application for permission to the Ministry of Land Infrastructure, Transport and Tourism at least 10 business days prior to the proposed operation.
- Remotely piloted aircraft may only be flown during the daytime.
- Remote pilots must maintain visual line of sight with their RPAS during operations.
- RPAS may not fly within 30 meters (98.4 ft) of people or private property.
- RPAS may not be flown over crowds or sites where large groups of people are gathered, such as concerts or sports events.
- RPAS may not be used to transport hazardous goods.
- RPAS may not drop objects while in flight, either intentionally or accidentally.

UK-based Skyports and Japanese trading company Kanematsu Corporation have announced a memorandum of understanding to expand their collaboration on advanced air mobility (AAM) in Japan. Skyports and Kanematsu are partnering on both vertiport and drone delivery solutions in Japan.

Urban Air Mobility in Japan is also in development. In April of 2021 the Ministry of Land, Infrastructure, Transport and Tourism in Japan announced the formation of the "Next Generation Aviation Mobility Planning Office" to handle regulation and issues surrounding aviation mobility. The Ministry also announced that it would regularize flight beyond visual line of sight (BVLOS) before 2023, the target date for introducing passenger drones and urban air mobility in Japan.

The Ministry also initiated an airspace restructuring plan for Japan's most travelled corridors and international operations in 2018, anticipating a need for Air Traffic Flow Control updates in 2025. No mention is made of a UTM or U-space equivalent, or even of a need for such architecture, but that will change if there is a demand for low level RPAS operations in controlled airspace.

Mexico

The Ministry of Communications and Transportation (SCT), through the General Directorate of Civil Aeronautics (DGAC), updated their drone regulations in April of 2015 with circular CO AV 23/10 R2. The DGAC has been replaced by the Civil Aviation Federal agency (AFAC). This circular establishes weight categories, as well as the usual operational limitations (no night flights, no flights near airports or sensitive areas, no flights over people and BVLOS, maintain minimum distances from people and objects on the ground, fly below 400 ft AGL, fly when visibility is good, no dropping of objects, and no flights in restricted areas). RPAs weighing less than 2 kg can be operated without authorization from the DGAC, but if operated commercially, they must have third-party liability insurance. RPAs heavier than 25 kg have more limitations, may only be operated with permission from DGAC (now AFAC), and the operator must have a pilot's license. Drones weighing more than 250 g must be registered with Mexican authorities. Only Mexican nationals can obtain a drone operator's license, unless there is a bilateral agreement between CAAs (such as the FAA) to permit foreign nationals to obtain a license.

No airspace integration efforts beyond the restrictions outlined above could be found in the public domain.

Netherlands

The Environment and Transport Inspectorate (ILT or ILENT) is the government agency charged with regulatory compliance and oversight of RPAS activities. The agency website is in Dutch and requires English translation. Any commercial RPAS operation requires an exemption from the ILT. The exemption rules require a certificate of registration, an airworthiness certificate, proof of insurance and competency of the pilot, an SMS for the organization, restrictions on where the RPAS can operate, and a two-person crew (pilot and camera operator or supervisor).

All drone operations conducted in EASA (European Union Aviation Safety Agency) member states must comply with current regulations, regardless of nationality. Non-EU residents are required to register as a drone operator with the National Aviation Authority of the first EU country in which they want to operate. If this is the Netherlands, the Netherlands Vehicle Authority (RDW) handles the registration.

As an EU Member State, the Netherlands complies with all relevant EU aviation regulations. The ILT website refers to Regulation (EU) 2018/1139 and Easy Access Rules for Unmanned Aircraft Systems Regulations (EU) 2109/947 and 2019/947. That will presumably include U-space implementation, as well as other airspace integration rules and means of compliance.

New Zealand

Part 101.201–215 (Remotely Piloted Aircraft, Control Line Model Aircraft, and Free Flight Model Aircraft) and Part 102 (Unmanned Aircraft Operator Certification) CAA Consolidation set forth the RPAS rules administered by the CAA of New Zealand. Part 101 is focused on aircraft weighing more than 15 kg and recreational modeling, although not specifically saying so, as the rules pertaining to approved persons or organizations, operations around aerodromes, airspace restrictions, visual line of sight restrictions, night operations, right of way rules, and aircraft mass limits generally contemplate aircraft operations in that category. Part 102 is specifically directed towards unmanned aircraft operator certification and, again, without directly so stating, is clearly intended for commercial RPAS operations. Generally, however, RPAs weighing less than 15 kg can fly in uncontrolled airspace below 400 ft AGL during daylight hours and within visual line of sight of the RPIC, in clear meteorological conditions, well clear of other aircraft, not within 4 km of an aerodrome, and outside of restricted areas, without obtaining permission of the CAA.

New Zealand initiated an Airspace Integration Trials Program to improve access to healthcare for communities across Aotearoa, New Zealand. The Airspace Integration Trials Program is led by the Ministry of Business, Innovation and Employment's (MBIA) Innovative Partnerships Program, supporting remotely piloted aircraft to integrate into the existing transport system and enabling businesses to test and develop advanced aerospace technologies in New Zealand.

This Program will involve determining route plans, certification plans, regulatory requirements, and linking up with other businesses. The New Zealand Government's objectives for the Program are to:

- Further position New Zealand as a location of choice for R&D in the development, testing, and certification of advanced unmanned aircraft and adjacent technologies;
- Realize the full potential for innovation within New Zealand's existing regulatory regimes;
- Generate evidence to inform medium- to long-term policy decisions on the integration of advanced unmanned aircraft, in line with the Government's vision for the unmanned aircraft sector set out in Taking Flight: An Aviation System for the Automated Age; and
- Increase awareness of the potential economic and social benefits arising from advanced unmanned aircraft applications and their integration.

New Zealand is also encouraging companies to invest in remotely piloted or autonomous air taxi technology, upper atmosphere RPA for remote sensing and environmental uses, agriculture, predator mitigation, and other innovative uses of drones.

Norway

Norway's CAA has jurisdiction over all RPAS activities. Relevant sections of the Aviation Act, as amended in 2016 and 2018, apply to all aircraft without a pilot on board. Commercial operations require permits from the CAA as well as the National Security Authority (if aerial photography is contemplated). All rules that apply to manned aircraft apply equally to RPAS. Recreational flying rules are similar to those contained in US regulations. Restrictions and limitations for all RPAS operations are divided into three categories: RPAS with a maximum takeoff mass (MTOM) of 2.5 kg and maximum speed of 60 kts (RO1); RPAS with MTOM of up to 25 kg and maximum speed of 80 kts (RO2); and RPAS with MTOM of more than 25 kg or maximum speed of 80 kts, or operated by turbine engine, or to be used for BVLOS operations higher than 120 m, or operating in controlled airspace higher than 120 m, or will be operated in urban airspace or over crowds of people (RO3). Each category has organizational and SMS requirements that are more stringent as the capabilities and mission of the RPAS increase in complexity.

The Norwegian approach to regulation is more comprehensive than most countries in nearly all respects, from aircraft certification and airworthiness, pilot qualifications, organizational safety culture, and to registration and marking. Beyond line of sight and extended visual line of sight operations may be performed if the license issued by the Norwegian CAA covers those types of operation. The common requirements found elsewhere also apply in Norway, such as lighting, night flights, flying in controlled airspace, and operating over disaster sites. The regulations also cover state operations (police, customs, fire, search, and rescue), but exempt those activities from the strict liability requirements imposed on all other commercial or recreational operations. Infringement fines may be imposed for violating any regulation.

Norway is a European Union Member State, and as such complies with the EU aviation regulations discussed in Chapter 2. Norway announced in February of 2020 that it will be the first Nordic country to implement a UTM system.

A technically advanced UTM solution will provide Avinor Air Navigation Services (Norway's ANSP) the means to accelerate safe integration of drones and help to further increase the use of this technology in Norwegian airspace. A company called "Frequentis" will join UTM technology provider Altitude Angel to deploy the UTM system at 18 airports across Norway.

The UTM system, which is being tested in a real-world environment at the first two airport towers before being rolled out nationwide, will support the country's future drone strategy. The UTM system will provide an operational overview of the airspace and will allow two-way communication between air traffic control (ATC) and drone operators.

The UTM project in Norway supports the Norwegian government's drone strategy for the increased safe use of drones, and will address the rising number of reported incidents and airspace violations. With the UTM solution, Norway's Air Navigation Service Provider will have the means to accelerate the safe integration of drones, evolving the commercial use of the technology to generate sustainable revenue streams.

Singapore

Singapore's aviation regulations are administered by the Civil Aviation Authority of Singapore (CAAS). The Statutes of the Republic of Singapore, the Air Navigation Act, Chapter 6, contains all relevant regulations. Division 4 – Special Powers and Prohibited Activities (added in 2018) – covers some aspects of unmanned aircraft operations. Air Navigation 101 – Unmanned Aircraft Operations – sets forth the UAS regulations.

UA registration was implemented in January of 2020. Any UA with a total weight of above 250 grams must be registered before operation. Registrants must be at least 16 years old at registration. From 2 April 2020, it is an offence to operate an unregistered UA in Singapore.

The RPAS regulations follow the standard model, categorizing UAs and their operating areas, pilot qualifications and licensing, operator and activity permits, airspace restrictions, privacy considerations, commercial operations, and penalties for non-compliance.

Two companies, OneSky and Nova Systems, recently completed trials for their proposed solution for the safe management of unmanned aircraft (UA) operations via an automated system. This is part of Nova Systems' work under the MOT/CAAS UAS Call-for-Proposal project, which seeks to develop technological capabilities for Unmanned Aircraft System Traffic Management (UTM) services. The trial will form the basis of a roadmap that helps safely and efficiently to usher in a large-scale rollout of UAS operations in Singapore.

The trial allowed for the testing of numerous advanced UTM services, some not seen in any other project: including flight planning and authorization, strategic deconfliction, conformance monitoring, real-time alerts, dynamic rerouting, constraint management, inter-USS communication and Remote Identification capabilities.

The trials bring to a close a two-year consortium project, led by Nova, and co-founded as part of the UAS Call-For-Proposals (CFP) by the Ministry of Transport (MOT) and Civil Aviation Authority of Singapore (CAAS). It marks a significant milestone in Singapore's journey towards a future driven by UAS technology, as private and government sector stakeholders look deeper into the possibility of the large-scale, integrated deployment of such technology in Singapore's unique urban environment. To conclude the trials, a UTM demonstration was held for CAAS on 10 March 2021.

Using a mix of live flights and simulations, the demonstration showcased key developments and findings from the team's work in UTM technologies over the past two years, in particular the handling of the mass deployment of drones flying Beyond the Visual Line of Sight. The technology developed and lessons learned presented numerous opportunities for Singapore's UAS industry for large-scale drone deployment capable of supporting a range of critical operations, including surveillance patrols and delivery of essential supplies. Under preidentified and monitored routes visualized through a UTM system, UAS operations are coordinated, regulated, and safely integrated into Singapore's urban airways, ensuring the safety and security of all citizens.

In Singapore and abroad, the immediate and long-term benefits of UAS technology have become increasingly evident. Growing demands for services such as last-mile delivery and mounting concerns over global issues like natural disasters and climate change have underscored the need to develop future-ready solutions to address the shortcomings of current urban infrastructures.

South Africa

The South African CAA (SACAA) makes and enforces the RPAS regulations, which came into effect in May and June of 2017. Civil Aviation Regulations Part 101 (RPAS) is divided into six subparts: general provisions; approval and registration; personnel licensing; RPAS operating certificate; RPAS operations; and maintenance.

The regulations do not apply to model and toy aircraft or autonomous unmanned aircraft that cannot be managed real-time during flight. Recreational aircraft can only be flown above one's own property or over property whose owner has granted permission. All commercial and non-profit operations are covered. Aircraft are grouped into four classes, with each class having subclasses, depending upon the type of operation (LOS, VLO, EVLOS), impact kinetic energy (in kJ), height (altitude), and MTOW. The Director of the SACAA must issue an RPAS Letter of Approval (RLA) before any RPAS can be flown commercially. All RPAS must be registered and marked. A remote pilot license is required, preceded by training and a skills test, and a logbook must be maintained. Operating certificates are issued by the SACAA. Operations are specifically controlled, like in other nations, and continued maintenance is required along with documentation.

Many of the operational requirements mimic manned aviation regulations (such as pre-flight checklists, duties of the pilot, right of way rules, time management, power reserves, radio communications, and the like). All in all, South Africa's RPAS regulations are comparatively comprehensive and leave little room for deviation from the rules.

Sweden

The Sweden Transport Agency has jurisdiction over all RPAS operations and provides an English translation of their Statute Book entitled TSFS 2009:88. All commercial UAS operations, research or tests for commercial operations, and all beyond line-of-sight operations require permission from the Agency. UAS activities are divided into four categories (1A, 1B, 2, and 3), based upon MTOW, impact kinetic energy, and whether the flight is to be BVLOS.

Organizational oversight, insurance, and registration and markings are required in all categories. Each category has its own specific set of requirements and limitations. The expected operational limitations for all categories are included (flights above crowds or urban areas, established safety zones, flights beyond the line of slight of the pilot, flights above 400 ft AGL, and flights in controlled airspace). Pilot qualifications, flight manuals, flight preparation and planning, meteorology, fuel and battery capacity, equipment requirements, reporting, communications, airworthiness certification, maintenance and documentation, quality control, and organizational integrity are all covered in detail. The Swedish UAS/RPAS regulations are as comprehensive as in any of the nations surveyed in this chapter, if not more so.

In September of 2021 the Swedish Transport Administration selected Unified Traffic Management (UTM) company Altitude Angel to supply its Guardian UTM Enterprise platform at Sweden's 2400 km² Drone Center in Västervik on Sweden's south-east coast. Altitude Angel's UTM platform will support Sweden's Positioning, Navigation, and Communications (PNK) project, an initiative to test, document, and evaluate the possibility and expediency of using the mobile network to position, navigate, and communicate with unmanned aerial vehicles beyond the visual line of sight, within Sweden's existing airspace rules and regulations.

Sweden is a European Union Member State and complies with the relevant EU aviation regulations in all respects. This project is similar to Norway's UTM project and both efforts will inform EASA's move towards full implementation of U-space by 2023.

United Kingdom

The UK CAA published the latest (seventh) edition of CAP722A & B in 2019. CAP722 is split into three separate documents: CAP722 "Unmanned Aircraft System Operations in UK Airspace – Guidance and Policy," CAP722A "Unmanned Aircraft System Operations in UK Airspace – Operating Safety Cases," and CAP722B "UAS Operations in UK Airspace – The UK Recognized Assessment Entity." These three documents are equal to or exceed in scope of any set of RPAS regulations enacted in any other nation.

CAP722, in its seven editions, was and is intended to provide guidance to those involved with the development, manufacture, or operation of UAS to ensure that operational authorizations may be obtained and the required standards and practices are met. Although intended for non-recreational users, the guidelines do overlap into the recreational category of activity. The document also outlines the safety, airworthiness, and operational requirements that must be met. These comprehensive regulations cover all of the expected topics that are found in other nations' UAS regulations.

CAP722A provides guidance for those involved in creating a safety case for supporting an application to the CAA for operational approval. This would involve a hazard and risk assessment that reduces or mitigates risk to a "Tolerable" and "As Low as Reasonably Practicable (ALARP)" level. This document is not strictly regulatory, but provides guidance and requirements for compliance with the regulations.

CAP722B is intended to harmonize UK regulations with European Union Regulations that took effect on 1 July 2020. The term "National Qualified Entity" has thus been changed to "Recognized Assessment Entity." The document describes the requirements, administrative processes, instructions, and guidance related to the Recognized Assessment Entity (RAE) scheme within the UK. The RAE is an organization approved by the CAA to submit reports and/or issue certificates on the CAA's behalf with regard to remote pilot competency. In other words, organizations that train and certify remote pilots, such as flight schools, and the list of approved RAEs is also published on the CAA's website.

The Air Traffic Management and Unmanned Aircraft Act 2021, Chapter 12, Part 3, covers six unmanned aircraft topics: powers of police officers and prison authorities; powers of police officers relating to ANO 2016; fixed penalties for certain offences relating to unmanned aircraft; amendment and enforcement regulations; disclosures of information; and Part 3: interpretation. None related directly to airspace integration issues.

The UK Civil Aviation Authority (CAA) Innovation Hub utilizes a "sandbox" research tool to help innovators maximize regulatory readiness for demonstration of their aviation systems by testing them in a safe environment and learning how they address regulatory challenges. The Regulatory Sandbox enables innovators to increase the regulatory readiness of innovative solutions that do not fit within the scope of existing regulations, permissions, and exemptions.

BVLOS operations in the UK are not explicitly prohibited or restricted by regulation, but do require authorization from the CAA. The UK CAA Regulatory Sandbox supported a Detect and Avoid (DAA) test solution comprising multiple technologies during flights within a Temporary Danger Area (TDA) to demonstrate its ability to operate safely in non-segregated airspace. The test flights gathered evidence on the functionality and reliability of the proposed DAA solution, supporting future plans for BVLOS operations in non-segregated airspace.

The Regulatory Sandbox will also support simulation and modeling to generate data that will inform the CAA on matters relating to:

- Regulatory gaps and challenges,
- Potential separation standards for electric air taxi operations,
- Efficacy of dedicated flight corridors,
- Design considerations for airspace solutions, and
- Airspace capacity with growing electric air taxi operations.

The early results from these studies have been promising, and suggest that it is feasible to support UAM operations while integrating with existing airspace users beyond what can realistically be accommodated today.

The UK UAM Consortium has been working on a concept of operations that addresses the regulatory challenges for UAS airspace integration. The next phase of the CONOPS development process will look towards the future air traffic management (ATM) procedures and ground infrastructure requirements that are necessary to support higher tempo, higher density flights while considering stakeholder and sustainability needs. This process will include stakeholder engagement, where industry experts will be engaged for their feedback on the concepts being developed.

Conclusion

In summary, this survey of UAS/RPAS regulations from a select few nations around the globe reveals many recurring themes. Common elements include aircraft weight and/or impact kinetic energy categories; speed and altitude restrictions; visual line of sight requirements; pilot qualifications and training; exclusions from restricted or sensitive areas; prohibitions from flying over assemblies of people; daytime/night-time restrictions; registration and marking; CAA approval for certain types of commercial operations; exemptions for recreational model aircraft; maximum weight and altitude restrictions; maintenance requirements; airworthiness standards; SMS and organizational criteria; licensing of pilots, entities, or operations; operational rules of the air; lighting of aircraft; dropping of objects and carrying hazardous materials; and third-party liability insurance requirements.

The ICAO RPAS Manual, the EASA U-Space initiative, the JARUS SORA documents, the FAA/NASA UTM partnership, all works in progress, will go a long way towards globally harmonizing basic UAS safety and operational criteria, while still allowing for ICAO Member States to promulgate their own variations of domestic UAS regulations. Although EASA and JARUS are directly concerned with European aviation safety, representatives from the FAA and other non-EU countries participate in their efforts to facilitate open dialog between the US and its European partners.

As UAS become more capable and approach manned aviation airworthiness and safety standards, global harmonization of the essential system integrity and operational rules becomes an essential component of the growth of the technology and its many humanitarian and utilitarian uses. The challenge of safely integrating RPAS of various sizes, weights, and capabilities into a non-segregated geographical zone, while respecting both national sovereignties and facilitating cross-border operations, is being aggressively met by civil aviation authorities and regional/global regulatory agencies, as well as supporting industries. The field is extraordinarily dynamic, with new innovations and developments being announced on a daily basis.

The task of developing standards and recommended practices to assist compliance with the evolving regulations, while simultaneously supporting regulatory authorities in promulgating those regulations, is the motivating force behind standards development organizations. The next chapter will offer an overview of the most active SDOs in the RPAS/UAS domain and will direct the reader to the more relevant efforts of each of those organizations.

References

Barnhart, K., Marshall, D., and Shappee, R. (2021). *Introduction to Unmanned Aircraft Systems*, 3e. Boca Raton, Florida: CRC Press.

La Franchi, P. (2021). AAM summit notes. September.

Spence, P. (2021). Chief Executive Officer and Director Aviation Safety, Civil Aviation Safety Authority. Remarks at Australian Advanced Air Mobility Summit, 1–2 September.

French ministries of transport and armed forces invite projects to accelerate U-space implementation. Unmanned Airspace website available at: https://www.unmannedairspace.info/latest-news-and-information/french-ministries-of-transport-and-armed-forces-invite-projects-to-accelerate-u-spaceimplementation/#:~:text= (accessed October 9, 2021).

6

The Role of Standards

In the US the FAA derives its power for making rules and regulations from the US Congress in the form of enabling legislation. In a country with such diversity of cultures, interests, geography, and human resources, the task of making laws and rules for every aspect of the US economy (at least those aspects that are established by the US Constitution) would be insurmountable. The same can be said for most developed and developing countries that boast some form of democratic governmental institutions. Consequently, governments will delegate the burden of overseeing and regulating, as necessary, the activities of its citizens to regulatory agencies. We have discussed some of those agencies and their efforts towards UAS integration in the previous chapters.

Regulatory agency rulemaking efforts are often supplemented or enhanced by published standards that are created by industry organizations and approved by the agency. EASA's various documents outlining Acceptable Means of Compliance and Guidance in support of European Union regulations are good examples of an agency's attempt to help the community it serves understand what the regulations mean, and how to comply with them.

Standards Development Organizations (SDOs) work with engineers, scientists, and other industry personnel to develop non-biased standards or specification documents that serve industry and protect the public. These participants can be and usually are private concerns, trade organizations, or professional societies. Standards providers are distributors of codes, standards, regulations, and guidance. They may also offer access to useful databases of standards. The supplier may or may not be the developer of the standards that are distributed. Memberships are open to all for a modest fee, and most business is conducted remotely due to the diversity of citizenship and residence of many members, particularly since face-to-face meetings were deferred due to the COVID pandemic.

These organizations are international professional societies made up of industry representatives, engineers, and subject matter experts from all over the world who provide advisory support to federal agencies such as the FAA, EASA, ICAO, and innumerable CAAs. They make recommendations that may become a formal rule by adoption or reference. In the US, once a consensus standard has been fully developed and ready for publication, the FAA may publish a Notice of Availability (NOA) in the Federal Register, wherein the agency declares that the standard is acceptable for certification of a product or procedure. The notice may invite public comments, which then must be adjudicated like any other

UAS Integration into Civil Airspace: Policy, Regulations and Strategy, First Edition. Douglas M. Marshall.
© 2022 John Wiley & Sons Ltd. Published 2022 by John Wiley & Sons Ltd.

proposed rule. In Europe, Engineering codes, standards, and regulations all serve to ensure the quality and safety of equipment, processes, and materials.

The three more prominent of those advisory organizations that are playing significant roles in the evolution of unmanned aviation are the Society of Automotive Engineers (SAE), the Radio Technical Commission for Aeronautics (RTCA), and the American Society of Testing and Materials (ASTM International). Other SDOs that are also contributing to UAS standards development are the American National Standards Institute (ANSI), the International Organization for Standardization (ISO), the Consumer Technology Association (CTA), the European Organization for Civil Aviation Equipment (EUROCAE), the Institute of Electrical and Electronic Engineers (IEEE), the Joint Authorities for Rulemaking of Unmanned Systems (JARUS), the Internet Engineering Task Force (IETF), and the International Telecommunications Union (ITU).

Standards supporting organizations (those that do not necessarily create standards, but provide subject matter expertise to the industry) include SESAR-JU (Single European Sky ATM Research Joint Undertaking), Global UTM Association (GUTMA), and the American Institute of Aeronautics and Astronautics (AIAA). In Europe EASA develops their own standards with non-regulatory, advisory documents branded "Acceptable Means of Compliance and Guidance Material," (AMC and GM), discussed more fully in Chapter 2. The American equivalent is a subagency of the US Department of Commerce, the National Institute of Science and Technology (NIST). The cellular industry has more recently become involved in unmanned aircraft technology, as cell phones and cellular connectivity have become integral components of remote identification and tracking and beyond visual line of sight (BVLOS) operations. Industry groups that also engage in standards development include 3GPP (3rd Generation Partnership Project) and GSMA (Groupe Speciale Mobile Association).

Aeronautical engineering codes are enforced by the CAAs and are critical to developing industry practices. While engineering *regulations*, such as those found in the US Federal Aviation Regulations or the European Union's Implementing Regulations, are government-defined practices designed to ensure the protection of the public, as well as uphold certain ethical standards for professional engineers. Engineering *standards* ensure that organizations and companies adhere to accepted professional practices, including construction techniques, maintenance of equipment, personnel safety, and documentation. These codes, standards, and regulations also address issues regarding system certification, personnel qualifications, airspace rules, enforcement, security, and many other areas of concern related to safety of the skies.

Manufacturing codes, standards, and regulations are generally designed to ensure the quality and safety of manufacturing processes and equipment, and aviation regulations are no exception. Manufacturing standards ensure that the equipment and processes used by manufacturers and factories are safe, reliable, and efficient. These standards are often voluntary guidelines, but can become mandatory by reference in the relevant regulations if adopted by the competent authority. Manufacturing regulations are defined by governments and usually involve legislation for controlling the practices of manufacturers that affect the environment, public health, or safety of workers. Aircraft manufacturers in the United States and European Union are required by law to produce aircraft that meet certain airworthiness and environmental emissions standards.

The FAA has supported and sponsored four international SDOs dedicated to developing standards and regulations for the manufacture and operation of unmanned aircraft.

RTCA Special Committee 203 Unmanned Aircraft Systems (SC-203) began developing minimum operational performance standards (MOPS) and minimum aviation system performance standards (MASPS) for UASs in 2004. "SC-203 products will help assure the safe, efficient and compatible operation of UAS with other vehicles operating within the NAS. SC-203 recommendations will be based on the premise that UAS and their operations will not have a negative impact on existing NAS users." SC-203's efforts ended in 2013, and a new Special Committee, SC-228, Minimum Operational Performance Standards for Unmanned Aircraft Systems, was created shortly thereafter, also in 2013. Its task is to develop MOPS for detect-and-avoid technology, as well as command-and-control (C2) data link MOPS seeking L-Band and C-Band solutions. This committee's work in part builds upon earlier standards published by the North Atlantic Treaty Organization, STANAG 4586, "Standard Interfaces of UAV Control System (UCS) for NATO Interoperability, 4th Ed."

ASTM F-38 Committee on Unmanned Aircraft Systems addresses issues related to design, performance, quality acceptance tests, operations, and safety monitoring for UASs. This committee consists of three subcommittees: F38.01 Airworthiness, F38.02 Flight Operations, and F38.03 Personnel Training, Qualification, and Certification. F38 collaborates with other ASTM committees and, as an international SDO, ASTM participates in standards development efforts around the world. ASTM stakeholders include manufacturers of UASs and their components, federal agencies, design professionals, professional societies, maintenance professionals, trade associations, financial organizations, and academia. On 6 October 2021 the F-38 committee's Working Group WK63418 released for ballot a draft UAS Traffic Management (UTM) UAS Service Supplier (USS) Interoperability specification. In addition, two other F-38 standards relating to UTM have been finalized for publication: F3411 *Standard Specification for Remote ID and Tracking* and F3201 *Standard Practice for Ensuring Dependability of Software Used in Unmanned Aircraft Systems* (UAS). Other relevant standards cited in the document are:

- ISO/IEC 27001:2013, Information Security Management;
- RTCA DO-278A, Software Integrity Assurance Considerations for Communication, Navigation, Surveillance, and Air Traffic Management (CNS/ATM) Systems;
- IETF RFC 3339, Date and Time on the Internet: Timestamps;
- IETF RFC 5905, Network Time Protocol;
- IETF RFC 6749, OAuth 2.0 Authorization Framework;
- IETF RFC 7519, JSON Web Token;
- EUROCAE ED-269, Minimum Operational Performance Standard for UAS Geo-Fencing; and
- ASD-STAN Pr4709-003, Geo-awareness requirements.

SAE's G-10U Unmanned Aircraft Aerospace Behavioral Engineering Technology Committee was established to generate pilot training recommendations for UAS civil operations and propose a standard that provides an approach to the development of training topics for pilots of UASs for use by operators, manufacturers, and regulators. The E-39 "Unmanned Air Vehicle Propulsion Systems" committee was chartered on 23 January 2018, with responsibility to develop and maintain standards for unmanned air vehicle propulsions systems and all their various components and subsystems. This committee released ARP5707 Pilot Training Recommendations for Unmanned Aircraft Systems (UAS) Civil Operations in 2016.

Additional international aviation organizations in Europe that exercise some level of regulatory powers include the European Organization for the Safety of Air Navigation (EUROCONTROL), EASA, and the European Organization for Civil Aviation Equipment (EUROCAE).

IEEE (Institute of Electrical and Electronic Engineers) "is the world's largest technical professional organization dedicated to advancing technology for the benefit of humanity. Among its many activities, IEEE sponsors standards development in a wide range of industries."

IEEE's Sponsoring Committee, COM/AccessCore-SC – Access and Core Networks Standards Committee, has recently chartered three subcommittees related to unmanned aircraft:

1. IEEE's Draft Standard for Drone Applications Framework (IEEE P1936.1) committee is developing: "A framework for support of drone applications is established in this standard. It specifies typical drone application classes and application scenarios and the required application execution environments. The general facility requirements of drone application are listed, including flight platform, flight control system, ground control station, payload, control link and data link, takeoff and landing system, etc. The drone safety and management requirements include airworthiness, airspace and air traffic requirements, qualification of operators, qualification of personnel, insurance, confidentiality and others. The general operation process is detailed. The operation results stipulate the operation record and operation report, including data classification, data collection and processing, data record and analysis and data reference format."

   The P1936.2: Photogrammetric Technical Standard of Civil Light and Small Unmanned Aircraft Systems for Overhead Transmission Line Engineering committee's scope is defined as: "The standard specifies the operational methods, accuracy indicators and technical requirements for the photogrammetry for light-small civil drone applications in power grid engineering surveys and design. The light and small civil drones in this standard refers to: (1) Fixed-wing UAV or multirotor UAV is applied as the flying platform; (2) Powered by battery or fuel; (3) The weight is between 0.25 kg and 25 kg without payload; (4) The maximum active radius is 15 km and the maximum operational altitude is 1 km."

2. Subcommittee P1920.2: Standard for Vehicle to Vehicle Communications for Unmanned Aircraft Systems is sponsored by VT/ITS Intelligent Transportation Systems and was organized in 2019. This committee's charge is:

   "Vehicle to Vehicle Communications (V2V) standard for Unmanned Aircraft Systems defines the protocol for exchanging information between the vehicles. The information exchange will facilitate beyond line of sight (BLOS) and beyond radio line of sight (BRLOS) communications. The information exchanged between the aircraft may be for the purpose of command, control, and navigation or for any application specific purpose."

3. Committee PN42.63: Recommended Practice for Unmanned Aerial Radiation Measurement Systems (UARaMS) project details are:

"This recommended practice establishes performance criteria and characterization techniques for radiation measurement systems incorporated onto unmanned aerial systems, or UARaMS. The recommended practice provides a means to accurately assess a UARaMS's effectiveness to either search/localize a radiological source or characterize/map radioactive contamination. It outlines measurement expectations, functionality characterization tests, and functionality needs based on available radiation response information, test results, and expected radiation fields at applicable heights above ground level (AGL). This recommended practice does not address individual unmanned aircraft system (UAS) performance or operations such as those items required for operation and control. The document looks at operational scenarios, detection needs and expectations, and environmental parameters that could be experienced during use, such as temperature changes, mechanical shock, and onboard vibration. For radiation, response vectors include those expected from distributed contamination and from point sources. The primary unmanned aerial vehicles (UAVs) of interest include those from Group 1 and Group 2 UAV designations."

NIST (National Institute of Standards and Technology) develops and disseminates the standards that allow technology to work seamlessly and business to operate smoothly.

"NIST supports accurate and compatible measurements by certifying and providing over 1300 Standard Reference Materials® with well-characterized composition or properties, or both. These materials are used to perform instrument calibrations in units as part of overall quality assurance programs, to verify the accuracy of specific measurements, and to support the development of new measurement methods.

Industry, academia, and government use NIST SRMs to facilitate commerce and trade and to advance research and development. Presently NIST SRMs are currently available for use in areas such as industrial materials production and analysis, environmental analysis, health measurements and basic measurements in science and metrology. NIST SRMs are also one mechanism for supporting measurement traceability in the United States.

Each NIST Standard Reference Material® is supplied with a Certificate of Analysis and a Materials Safety Data Sheet, when applicable. In addition, NIST has published many articles and practice guides that describe the development, analysis, and use of SRMs.

The SRM Program references a number of definitions in connection with the production, certification, and use of its SRMs and RMs. Certain definitions, adopted for SRM use, are derived from international guides and standards on reference materials and measurements while others have been developed by the SRM Program to describe those activities unique to NIST operations.

An NTRM (NIST Traceable Reference Material) is a commercially produced reference material with a well-defined traceability linkage to existing NIST standards. This traceability linkage is established via criteria and protocols defined by NIST. Commercial reference materials producers may affix the NTRM trademark to materials produced according to these criteria and protocols."

EUROCONTROL is an intergovernmental organization that acts as the core element of air traffic control services across Europe and is dedicated to harmonizing and integrating air navigation services in Europe and creating a uniform air traffic management system for civil and military users. The agency accomplishes this by coordinating the efforts of air traffic controllers and air navigation providers to improve overall performance and safety. The organization is headquartered in Brussels and has 38 member states. The European Commission created the SESAR initiative in 2001 and has since delegated portions of the underlying regulatory responsibility to EUROCONTROL. As discussed in Chapter 2, EASA was established as an agency of the European Union in 2003 and has regulatory responsibility in the domain of civilian aviation safety, thereby assuming the functions formerly performed by the Joint Aviation Authorities (JAA). In contrast to JAA's role, EASA has been granted legal regulatory authority by the European Commission, which includes enforcement powers. EASA is responsible for airworthiness and environmental certification of aeronautical products manufactured, maintained, or used by persons under the regulatory oversight of European Union member states, as well as airspace management and development, particularly the U-space project.

EUROCAE (European Organisation for Civil Aviation Equipment) reports to EASA, although it was created many years before EASA was formed, and is exclusively chartered with aviation standardization (with reference to airborne and ground systems and equipment). EUROCAE is not a regulatory body, but its standards and related documents are required for use in the regulation of aviation equipment and systems. Its membership is made up of equipment and airframe manufacturers, regulators, European and International CAAs, air navigation service providers, airlines, airports, and other airspace users. EUROCAE's Working Group 73 is devoted to the development of products intended to help assure the safe, efficient, and compatible operation of UASs with other vehicles operating within non-segregated airspace. WG-73 makes recommendations to EUROCAE with the expectation that those recommendations will be passed on to EASA.

More recently, EUROCAE WG-105, "UAS," was tasked with developing standards and guidance documents for the safe operation of UAS in all types of airspace, at all times, and for all types of operations. WG-105 originally consisted of six Focus Teams working in specific focus areas, identified as Sub Groups (SG): UAS Traffic Management (UAS); Command, Control, Communications (C3); Detect and Avoid (DAA); Design and Airworthiness Standards; Specific Operations Risk Assessment (SORA); and Enhanced RPAS Automation (ERA). Subsequently, the number of WG-105 subgroups was expanded to include new subgroups:

- SG-0 (Steering Committee);
- SG-10 (Detect and Avoid Focus Team);
- SG-11 (DAA against conflicting traffic for UAS operating under IFR in Class A-C airspaces);
- SG-12 (DAA against conflicting traffic for UAS operating under IFR and VFR in all airspace classes);
- SG-13 (DAA for UAS operating in VLL);
- SG-20 (C3 and Security Focus Team);
- Sg-21 (RPAS C2 Data link);
- SG-22 (Spectrum);

- SG-23 (Security);
- SG-30 (WG-105 UTM Focus Team);
- SG-31 (UTM-General);
- SG-32 (UTM-E-Identification);
- SG-33 (UTM Geo-fencing);
- SG-40 (WG-105 Design and Airworthiness Focus Team);
- SG-41 (RPAS System Safety Assessment Criteria);
- SG-42 (Remote Pilot Stations);
- SG-50 (WG-105 ERA Focus Team);
- SG-51 (ERA – Automatic Take-off and Landing;
- SG-52 (ERA – Automatic Taxiing);
- SG-53 (ERA – Automatic and Emergency Recovery);
- SG-60 (WG-105 SORA Focus Team);
- SG-60 (SORA);
- SG-62 (GNSS for UAS);
- SG-63 (Automatic protection function for UAS.

WG-105 works in coordination with RTCA SC-228 for UAS.

ANSI (American National Standards Institute) oversees standards and conformity assessment activities in the United States.

> "ANSI's mission is to enhance both the global competitiveness of US business and the US quality of life by promoting and facilitating voluntary consensus standards and conformity assessment systems, and safeguarding their integrity. Encompassing nearly every industry, the institute represents the diverse interests of more than 270,000 companies and organizations, and 30 million professionals worldwide. The ANSI UASSC's mission is to coordinate and accelerate the development of the standards and conformity assessment programs needed to facilitate the safe integration of unmanned aircraft systems (UAS) into the national airspace system of the United States. The collaborative is also focused on international coordination and adaptability. The overarching goal is to foster the growth of the UAS market, with emphasis on civil, commercial, and public safety applications. The aim is to describe the current and desired future standardization landscape, articulate standardization needs, drive coordinated standards activity, minimize duplication of effort, and inform resource allocation for standards participation."

The ANSI UASSC Roadmap 2.0 Workplan describes ANSI's goal of tracking standardization gaps in the work of other SDOs. Their latest gap analysis highlighted eleven areas that bear further analysis and development. All but one (implementing UAS for hydrocarbon pipeline inspections) have some relevance to UAS airspace integration.

ISO (International Organization for Standardization is a worldwide federation of national standards bodies (ISO member bodies). International organizations, governmental and non-governmental, in liaison with ISO, also take part in the work. ISO collaborates closely with the International Electrotechnical Commission (IEC) on all matters of electrotechnical standardization. In total, ISO collaborates with over 700 international, regional, and national organizations. These organizations take part in the standards development process as well as sharing expertise and best practices.

The work of preparing International Standards is normally carried out through ISO technical committees. The ISO TC 20 Unmanned Aircraft Systems, Subcommittee SC 16, UAS Traffic Management has produced several drafts of proposed standards for UTM, including a UTM framework, UTM functional structure, methodology for determining UTM functions, classification of UTM functionalities, traffic management functions, ATM coordination functions, operation support functions, reporting, and supplemental data supply functions. This is the first standard in the ISO series 23629, covering functional architecture, semantics, and a range of UTM services and requirements for the service providers.

The mission of ISO Technical Committee TC 20 is "the Standardization in the field of unmanned aircraft systems (UAS) including, but not limited to, classification, design, manufacture, operation (including maintenance) and safety management of UAS operations."

In September 2021 the same Sub-Committee SC 16 of ISO Technical Committee TC 20 voted unanimously on the final draft of the new International Standard ISO 23629-7 on UAS Traffic Management (UTM), Part 7: Data model for spatial data. The scope of the document states:

> "This document specifies the data model that is related to various spatial information for common use between the UAS service provider and the system for operation control, e.g., UTM. This document specifies the names of the items for the data model, while the communication architecture and responsibilities of actors to define the items are not included."

This same technical committee is developing a number of other standards pertaining to unmanned aircraft systems, some still works in progress, others out for ballot among committee members.

AIAA (American Institute of Aeronautics and Astronautics) is accredited by ANSI and manages a wide range of national aerospace standards publications and activities. AIAA was selected as the administrator of the ISO TC20/SC16 committee, discussed above. Standards development groups within AIAA focus on UAS components, evolving technology certification, safety, and operations. A UAS Standards Working Group was tasked in 2017 to explore gaps in standards development, and eventually became involved in ANSI's parallel effort. AIAA has undertaken some standards development in the UAS sector, but its major contribution is the support of the ISO effort and to host regular programs and educational events where participants present papers on a wide variety of topics related to unmanned aircraft.

GUTMA (Global UTM Association). "GUTMA aims to foster the safe, secure, and efficient integration of drones in global airspace systems through a supporting mission that accelerates transparent implementation of interoperable UTM systems." The organization's goals are to empower a harmonized global digital UTM ecosystem, and to enable more than the ATM Trusted source for UTM data and ideas (i.e. Knowledge Repository). GUTMA does not make regulations and standards, but acts as a support system or resource. While it is not a standards making organization, it is included in this chapter as a reference for additional information on UTM developments. The organization released a document entitled "UAS Traffic Management Architecture" in 2017.

The document describes an overall high-level UTM architecture. It considers all types of UAS operations (VLOS, EVLOS, and BVLOS), and covers the needs of both RPAS (piloted) and autonomous unmanned aircraft. The scope of the UTM architecture described in the document focuses on a UTM solution for UAS operations in very-low-level airspace, and addresses the requirements for all phases of the flight.

The document answers the following questions:

- What is a UTM?
- Who are the stakeholders?
- What are the stakeholders' interests and interactions with the system?
- How can users interact with the system?
- What are the functions of a UTM?

The document offers a helpful comparison between "UTM Concept" and "UTM System."

> "The **UTM Concept** is a complex system in which several stakeholders contribute to ensure the required safety level of UAS operations. For this reason, UTM is defined as a system of stakeholders and technical systems collaborating in certain interactions, and according to certain regulations, to maintain safe separation of unmanned aircraft, between themselves and from ATM users, at very low level, and to provide an efficient and orderly flow of traffic."

On the other hand, a "**UTM System** is a concrete technical implementation comprising software, the necessary infrastructure for running the software, and the drones themselves, all contributing to the achievement of UTM. The system provides distinct services through public or restricted standard interfaces. Individual services are provided at distinct levels of quality depending on the situation or regulation (from best effort to high-integrity, low-latency). Different systems providing similar/equivalent/interdependent/interfering services within the same area of effect are required to collaborate.

The UTM concept covers all type of UAS operations in very-low-level airspace, in all categories, ranging from simple remotely piloted aircraft systems to complex autonomous operations and beyond."

The GUTMA Annual Report 2021 lists several activities related to UTM development:

- GUTMA – GSMA COLLABORATION – THE AERIAL CONNECTIVITY JOINT ACTIVITY (ACJA)
- "MAP OF UTM IMPLEMENTATION"
- WORK ON STANDARDS
- GUTMA MOVED FROM MICROSOFT TO GOOGLE GSUITE ENVIRONMENT

Conclusion

The organizations discussed in this chapter, as well as others not named, provide valuable support to industry, governments, and regulators as they work through the maze of challenges and emerging technologies to achieve the common goal of a global air traffic management system that will accommodate commercial RPAS operations, eVTOL air taxis, piloted, remotely piloted, and autonomous aircraft at the lower levels, and high-altitude long-endurance aircraft operating at or above commercial flight levels. Those aircraft in the latter category will have to transition through multiple classes of airspace to get to Class A and unlimited Class E. Weather balloons do that now, with strict restrictions, but there are countries and companies developing hypersonic aircraft that can fly at FL600 or above. A new airspace architecture will have to be developed to meet these diverse operational needs, and the many SDOs comprised of teams of visionary and insightful subject matter experts will help pave the path towards that goal.

For the reader, joining any one (or more) of these organizations can prove to be a rewarding experience, exposing one to exciting and challenging work with dedicated people from around the world. Bringing something to the table in these organizations often leads to new opportunities for those who are so motivated.

References

AIAA documents and standards available at: https://www.sae.org.

ANSI standards available at: https://www.ansi.org.

ASTM standards available at: https://www.astm.org/Standards/F3411.htm and https://www.astm.org/Standards/F3201.htm.

EUROCAE standards available at: https://eurocae.net.

GUTMA UAS Traffic Management Architecture available at: https://www.gutma.org/docs/Global_UTM_Architecture_V1.pdf.

GUTMA Annual Report 2021 available at: https://gutma.org/annual-reports/ (accessed October 10, 2021).

ISO standard is available on the ISO website https://lnkd.in/dVqXjwHj.

JARUS documents available at: https://rpas-regulations.com/community-info/jarus-publications.

NIST standards available at: https://www.nist.gov/standards.

RTCA standards available at: https://www.rtca.org.

SAE documents available at: https://www.sae.org.

7

The Technology

"Form follows function," is a concept or methodology in architecture attributed to Louis Sullivan (1856–1924) in his 1896 essay, "The Tall Office Building Artistically Considered," published the same year that the Prudential Guaranty Building was constructed in Buffalo, New York. "Sullivan's legacy – besides instilling ideas in his young apprentice, Frank Lloyd Wright (1867–1959) – was to document a design philosophy for multiuse buildings. Sullivan put his beliefs into words, ideas that continue to be discussed and debated today" (Craven 2019).

Although commonly used to describe architectural design concepts, the fundamental idea works very well in other engineering or technology fields, and is useful to keep in mind when discussing the design and implementation of UAS airspace integration architectures such as U-space, UTM, and AAM/UAM.

The SESAR Blueprint states that, over time, U-space services will evolve as the level of automation of drones increase, and advanced forms of interaction with the environment are enabled (including manned and unmanned aircraft) mainly through digital information and data exchanges. Whether the task is designing and building a skyscraper or designing and implementing an airspace construct around that building that allows safe operations of UAS in an urban environment, the core concept is the same. The developers must determine what function the airspace is to achieve or serve, and in what form can it be developed and implemented, all the while conforming to existing or evolving regulations.

The JARUS SORA Special Operations Risk Assessment tool (discussed in Chapter 2) is a complex and comprehensive methodology for analyzing and quantifying risk. A similar process (although not as lengthy as SORA) has been required by the FAA for many of its regulatory and safety initiatives. From FAA Order 8000.369 C:

> 1. This order establishes the Safety Management System (SMS) policy and requirements for the Federal Aviation Administration (FAA). The requirements contained within this document are intended to help FAA organizations incorporate SMS and/or International Civil Aviation Organization (ICAO) State Safety Program (SSP) requirements into their organizations. FAA organization SMSs work together to form the overall FAA SMS. Specifically, this order:
>
>   a. Furthers safety management by evolving to a more process-oriented system safety approach with an emphasis on Safety Risk Management (SRM) and Safety Assurance processes.

UAS Integration into Civil Airspace: Policy, Regulations and Strategy, First Edition. Douglas M. Marshall.
© 2022 John Wiley & Sons Ltd. Published 2022 by John Wiley & Sons Ltd.

b. Sets forth basic management principles to guide the FAA in safety management and safety oversight activities.

c. Requires adopting a common approach to implementing and maturing an integrated SMS, including fostering a positive safety culture and other attributes as applicable.

d. Defines the roles and responsibilities of the FAA organizations, FAA SMS Executive Council, and FAA SMS Committee regarding safety management.

2. This order applies to the following Lines of Business (LOBs) and Staff Offices: Air Traffic Organization (ATO), Aviation Safety (AVS), Airports (ARP), Commercial Space Transportation (AST), Next Generation Air Transportation System (ANG), and Security and Hazardous Materials Safety (ASH). This order is written to allow for application to other FAA organizations as deemed appropriate by the Administrator.

"SMS introduces an evolutionary process in system safety and safety management. SMS is a structured process that obligates organizations to manage safety with the same level of priority that other core business processes are managed. This applies to both internal (FAA) and external aviation industry organizations (Operator and Product Service Provider)."

Developers and regulators in the US are required to follow this process as they work to design and implement the UTM CONOPS. Similar requirements can be found in the EASA U-space program (JARUS-SORA) and elsewhere.

The following paragraphs incorporate some of the components of the SORA process, as well as the FAA's SMS, albeit greatly simplified, and are intended to introduce the reader to the tools that may be available to perform a risk analysis of a new or modified aviation traffic management system.

As we have seen in the preceding chapters, there are several domains in the UAS integration realm that must be considered as the programs progress through design and implementation. At the regulatory and operational level, the more obvious categories could be initial airworthiness, continuing airworthiness, remote pilot qualifications and licenses, aircraft certification and operations, operator qualifications, rules of the air, ATM/ANSPs, and aerodromes, among others. However, this short list does not tell the whole story.

As previously noted, the exponential growth of drones globally (80,000 registered drones in the US in 2017, 869,428 as of August 2021, and there may be as many as 1.2 million, accounting for those that are not registered or are not required to be registered by regulation) has stressed the regulators and governments who are attempting to serve the needs of a new technology sector while maintaining the safety levels and system integrity of the legacy infrastructure.

Integration of RPAS into the airspace between 500 ft and 60,000 ft as either IFR or VFR is challenging due to the fact that RPAS will have to fit into the existing ATM environment and adapt accordingly. Many RPAS technical characteristics such as C2 latency and see and avoid capability were not previously at issue within the manned aviation environment, simply because a pilot is onboard the aircraft and capable (in theory) of handling these duties in a safe and timely manner. In addition, these human competences typically have not been translated into system performance requirements, as they were categorized under "good airmanship" for see and avoid, or simply not addressed at all (EUROCONTROL RPAS ATM CONOPs). That gap is being closed by development of packages of on-board see and avoid technology (combining real-time visual and motion sensing with other aids to detect intruding aircraft of objects), with the ultimate goal being to replicate or even improve upon the human capability to detect and analyze the potential threat of a moving or stationary object.

One of the major factors in designing or adapting an integrated airspace architecture is the human/machine interface problem. A thorough functional analysis of any element of a system that includes both humans and some form of automation should address these eight categories of human–machine interface and how they might intersect in designing a complex system:

- Human-in-the-loop
- Human-on-the-loop
- Human-over-the-loop
- Human-out-of-the-loop
- Automation-out-of-the-loop
- Automation-over-the-loop
- Automation-on-the-loop
- Automation-in-the-loop

This iterative process is a pervasive element in the design of manned aircraft control and navigation software and hardware. It should be no less so as regulators, developers, scientists, and engineers work to construct an airspace design that can accommodate the myriad uses of the airspace that are being proposed now and will be for the foreseeable future.

The task at the top level is to define an aircraft and ATM function list, identify the functions that will or could involve automation, locate the function on an authority versus responsibility matrix, carefully deconstruct all components relative to each function, then validate the result against relevant regulations, standards, guidance materials, and means of compliance.

Such an analysis can begin by considering the following functions for just a UAS/RPAS, without consideration for the ATM system in which it operates:

The core functions for the UAS pilot are to:

Aviate (fly the airplane): This refers to controlling the flight and ground paths; air to ground transitions; command and control (C2) functions between the ground control station (GCS) and the UAS; and control of the UAS subsystems.

Navigate (know where you are going): This involves conveying the navigation state (position, airspeed, altitude); establish the navigation intent; generate a navigation command; and determine the status of the navigation command.

Communicate (talk to controllers): This requires broadcasting critical information to ATC and other aircraft and receiving information from the same sources.

Mitigate (react if a hazard is recognized): The last function is to avoid collisions; avoid adverse environmental conditions; and manage contingencies.

Integrating the functions of the notional automated ATM system contemplated by the developers of U-space, UTM, AAM, or UAM into the overall "system of systems" (UAS and the supporting airspace traffic management infrastructure) complicates the risk analysis by an order of magnitude.

To begin with, it is necessary to identify and define the functions or tasks involved. Is there a standard approach to analyzing the safety requirements of increasingly autonomous aviation systems? What are the domains or themes that characterize the system under consideration? At the top level, the domain could be the structure and use of a national airspace system itself, or, on an even higher level, the global air transportation system (as overseen by ICAO, working in concert with national civil aviation authorities). Within that overarching domain are the competent regulatory authorities, the governmental bodies that are charged with ensuring the safety of the common asset (the air above the surface), and the user community.

As the analyst works down through the domains, from the top (airspace) to the lowest level (perhaps the design and integrity of a component of a UAS or an ATM structure, such as a computer chip or even further down to the human beings and their design processes and management structures that design the chips, etc.), the temptation to get "down in the weeds" could lead to "mission creep," or going off on tangents that divert from the original scope of the analysis. However, a thorough functional decomposition of a system as complex and fraught with potential single point failures as UTM cannot overlook even the finest detail or component. Which leads to the next set of questions:

What are the unintended consequences of transitioning from human-in-or-on-the-loop control systems (whether in the air or on the ground) to automated or semi- or fully autonomous interfaces? Will the automation associated with the function expose the users of the airspace to misuse by rogue actors? Can the system be disrupted at any point by a hacker, environmental issues, or component failure in a manner that the entire system ceases to function, even for a moment? System-wide air traffic control failures (or shutdowns) are rare, but they do happen.

A graphic example of this type of system breakdown is the catastrophic failure of the space shuttle *Challenger* (OV-099, STS-51-L) in 1986, resulting in loss of the spacecraft and its entire seven-person crew. Although this was not an automation failure, after an intense inquiry was conducted the investigators determined that the failure of a rubber "O-ring," made by the solid-rocket booster's manufacturer Morton Thiokol, was the immediate proximate cause of the disaster. This was due to exposure to unseasonal and extreme cold temperatures at the site of the launch pad in Florida on the morning of the launch. At the launch site, the rocket fuel booster segments were assembled vertically. Field joints containing rubber O-ring seals were installed between each fuel segment. The official investigation revealed that the O-ring was manufactured in compliance with the specifications dictated by NASA, but the O-rings were never tested in extreme cold. On the morning of the launch, the cold rubber became stiff, failing to fully seal the joint.

The suspect component (a very small part of a large and hugely complex space vehicle) was not designed to withstand temperatures below a certain level because such an exposure could (and did) cause the O-ring to crack or split and allow an inflammable plume of exhaust gas to leak out, with predictable results. The hot gases coated the hull of the cold external tank filled with liquid oxygen and hydrogen until the tank ruptured. The investigation also revealed that a Morton Thiokol engineer on site, who was a member of the launch team, warned the leadership of that possibility, yet the launch was approved nevertheless. What emerged from the Rogers Commission's investigation was a shocking pattern of assumptions by NASA and Morton Thiokol engineers and managers that the vehicle could survive minor mishaps and could be pushed beyond its design capabilities. The Commission heard testimony from several engineers who had expressed concerns about the reliability of the seals for at least two years, and who had warned their superiors about a possible failure just the night before the *Challenger* was launched. That NASA ignored these warnings, resulting in a tragic loss of life and the temporary shutdown of NASA's space shuttle program, was the subject of a number of books and papers describing the organizational breakdown that allowed the disaster to happen. An excellent treatment of this classic example of a single point failure in a complex system is Diane Vaughan's book *The Challenger Launch Decision: Risky Technology, Culture and Deviance at NASA* (Vaughan 2016).

The next question to consider is the range of possible unintended consequences. The example just discussed demonstrates the most undesirable and perhaps preventable consequence (death or injury of human beings). The list of possible consequences could range from the worst (death of a person, disruption of a system, failure of a company, temporary shutdown of one aspect of the system) to the least intrusive (timely and uncomplicated replacement of the failed component or person). A large part of this analysis is determining what checks or bounds (architecture, software, personnel, etc.) must be established to limit unintended consequences. Commercial aircraft, by regulation, have redundancies built into their systems, so that a failure of one critical element of the system is backed up by a second or third component that takes over when the primary component fails. The task then is to determine the protocols for how that transition takes place – is it manually initiated by the pilot, air traffic controller, or operator, or does the system make the decision autonomously? What are the criteria for making that determination at the design stage or even in the operational phase? And where does the human element factor in to the analysis? At some fundamental level, it can be argued that every failure of a system or device can ultimately be traced back to human error (at any stage, from concept, to design, to testing, to production, to implementation, to training, to inspection and maintenance, and to management oversight).

Also to be considered is the cultural and organizational aspect. How invested is the organization in the trend towards automation, or the public that is affected by the technical evolution? Will there ever be a time when people will be comfortable with boarding a commercial aircraft that can accommodate hundreds of passengers, where that aircraft is being piloted by an entirely autonomous system? Many visionaries and entrepreneurs seem to have proposed that this is in the realm of possibility in the foreseeable future, and claim that the current technology can support that concept of operations (recall that Fred Smith, founder of package delivery carrier FedEx, declared over 10 years ago that current aircraft systems had the technology and capability to conduct such operations then, and presumably more so now). The *Challenger* disaster might have been averted if the culture at NASA had not been driven by pressure from the political class in Washington to move faster in developing manned space exploration vehicles, with ever-dwindling resources, thereby putting organizational comfort ahead of safety. The rush to score a success in the program clouded the judgment of the decision makers, again with predictable results.

Ideally, the risks and safety benefits should also be carefully deconstructed to ensure completeness of the analysis. What is the Target Level of Safety that the system is trying (or required by regulation) to achieve? Should an operational risk assessment process like

SORA or its equivalent be employed to quantify the likelihood of failure and the consequences of the automation not being employed?

Also, in the context of airspace integration, what are the allowable meteorological conditions and flight rules for the activation of the system, and what elements are required to ensure compliance with the regulations? Is U-space or UTM intended to be an all-weather, IFR operation (remotely piloted autonomous or semi-autonomous aircraft are flying blind, so to speak, without a pilot on board, but that does not account for manned aircraft operating in the same airspace)? The FAA/NASA UTM CONOPs and Europe's U-space notional architectures are founded upon the idea of system interoperability where all participants have equal access to critical information and airspace configurations are discovered and activated autonomously. Ultimately, human intervention may not be necessary, or even permitted.

Other questions that should be addressed in a comprehensive risk analysis are:

- What are the risks to people on the ground, and can the operation be safely activated over people?
- What safety mitigations are required?
- What are the intended/contemplated operations of other aircraft in the airspace (commercial carriers or private individuals; passengers vs. cargo)?
- Does the failure of any element of the system prevent a safe recovery or emergency landing of an aircraft in an acceptable manner (is it a catastrophic failure)? How can the failure be identified, and is there a regulatory requirement for a "fail safe" system? Are there more than one potential single points of failure?
- What are the consequences of a series of single point failures, any one of which could create a hazard, but in combination may result in a catastrophic outcome?
- What are the safety consequences of *not* implementing the system (compared to the status quo)?
- What is the level of reliability of the human remote pilot or air traffic controller performing the system's function? How can this be determined? Is there enough data to support an analysis? What tools are available to conduct this type of analysis?

The Role of Automation is another domain to be considered:

- How can the function of UTM/U-Space be performed without the proposed automation?
- What is missing from the UTM/U-space CONOPS, if anything?
- What needs to be invented to close the circle for implementation, if anything?
- Does automation merely assist the human in performing the function, or does the automation execute the function without human intervention?
- What is the nature of the safety monitor (e.g. human vs. system)?
- What is the nature of the fallback or backup to the system (e.g. human vs. system)?
- What is the role of the system (primary, blended, or safety enhancing)?
- Consider the implications of human-in-the-loop, human-on-the-loop, human-over the-loop, and human-out-of-the-loop in system design.

- A responsibility assignment matrix (RACI) or linear responsibility chart (LRC) that describes the participation by various roles in completing tasks or deliverables for a project or business process can be useful.

System complexity and maturity is another area of functional analysis. At first, the maturity level of any system under development must be determined. Is it merely a notional concept, is it still at a testing stage (like the NASA/FAA UTM activities, which were reportedly at the final TCL4 demonstration level in 2019), or is the system a functioning reality? In the NASA/FAA example, TCL4 was defined by the NASA UTM CONOPs to demonstrate: (1) beyond-visual-line-of-sight operations; (2) urban environments, higher densities; (3) autonomous vehicle-to-vehicle (V2V) capabilities, internet-connected systems; and (4) large-scale contingency mitigation. That phase of research and testing has concluded and NASA has moved on to the next level. And what is the next level?

What is the maturity level of the system components in terms of technical readiness levels (TRLs)? Is the network dependent upon 4 G or 5 G connectivity, for example? If mobile phone applications are to be the portal for remote pilots or operators to gain access to flight authorizations for UTM airspace, will the federated network be compatible with all versions of smart phones or tablets, or just the latest? What happens when the operating systems of those devices are updated by the manufacturers? Will the network have to be upgraded or updated as well? How will the system administrators communicate the need for software updates to its users, and who will provide the updates?

How can the service history of analogous systems (if any) be leveraged? Are there any analogous systems other than legacy ATM systems? Is there any existing system that is as dynamic as airspace? As the analysts drill down into the minutia of the components of the system, what mechanism exists to identify and leverage analogous systems or their core features? Are there reliability data available for each component?

The analyst must define the complexity of the system performing the function. The U-space/UTM architecture as envisioned by its proponents is a high-level conceptual framework that is made up of many interconnected components. Each of those components has its own architecture. The interdependency of each element of the structure generates complexity, which compels further in-depth examination to appreciate the degree of complexity of the overall system.

Are there performance monitors in the system that are designed to bound system performance? Are there regulatory requirements or standards in place that define the functionality and reliability of those monitors? What mitigations should be in place to account for the failure of the monitor, or pilot/operator error in failing to respond to an alert of an impending performance anomaly? Can this function be entirely automated, taking the response out of the hands of the human in- or on- the-loop?

How are safety-enhancing behaviors/attributes leveraged? The big question is whether the interdependencies among the system elements or domains have been fully accounted for in the design, or are there gaps (known unknowns versus unknown unknowns, with apologies to Donald Rumsfeld)?

Last, operational mission benefits need to be considered. How will employment of the system benefit the aviation community, and how will the broader community (or society at large) benefit from the employment of the system? And, does the system change the barriers to entry for operators and remote pilots? In other words, are training requirements and costs going to be affected by the implementation of the new system? A recent example of this challenge is the loss of two Boeing 737 Max aircraft in a short period of time, resulting in the deaths of 346 people. The issue now pending in the courts and the regulatory oversight and enforcement process is whether a new software design for the flight management system (MCAS) required additional training (and attendant costs to the airlines affected), and whether Boeing should have published or made available new training protocols to teach pilots on how to respond to certain anomalies in aircraft controllability. While investigations are still ongoing, the similarities in the two events raise the specter of improper relationships between the manufacturer and the FAA, leading to a lack of proper oversight over material design changes that impacted the airworthiness certificate of that aircraft model. Whether that human error falls at the feet of Boeing or the FAA (or both) has yet to be determined. But the reality is that the two disasters would not have happened but for human beings making bad decisions. The proof is in determining where in the chain of causation the human factor intervened to create the problem.

Conclusion

Introducing a new technology into an existing infrastructure presents many challenges to the developers, regulators, users, and the general population. In the case of UAS/RPAS airspace integration, three mantras from the manned aviation community have hovered over the process at all levels and with all stakeholders for years.

First, there should be no reduction in safety as a consequence of permitting new types of aircraft and operations into the existing air traffic management system. Second, no new equipment should be mandated on manned aircraft. And third, no new airspace should be created.

The first concern is self-evident, and warrants no further discussion or debate. While the mere addition of over 800 000 drones into the US airspace clearly presents the potential for a dimished level of safety, the FAA has taken steps to mitigate that threat by the development of comprehensive rules for the operation of remotely piloted aircraft. The same can be said for the European Union and other nations around the world. In the US, the FAA has also created the LAANC airspace management model to allow UAS operations in the airport environments of over 700 airports. Remote Identification and Operations Over People rules were also enacted to mitigate the impact of UAS sharing low level airspace with manned aviation.

The second mantra, no new equipment on manned aircraft, has by and large been honored in the US. However, the FAA now requires ADS-B Out capability in the continental United States, in the ADS-B rule airspace designated by FAR 91.225, but that took years to implement because of aggressive push back from the general aviation community. ADS-B is required to operate in: Class A, B, and C airspace; Class E airspace at or above 10 000 ft MSL, excluding airspace at and below 2500 ft AGL; within 30 nautical miles of a Class B primary airport (the Mode C veil); above the ceiling and within the lateral boundaries of

Class B or Class C airspace up to 10 000 ft; and in Class E airspace over the Gulf of Mexico, at and above 3000 ft MSL, within 12 nm of the US coast.

When UTM airspace is implemented, there may be additional equipage requirements for manned aircraft to operate in unsegregated UTM airspace, but the details of any additional requirements have not emerged. It may well be that Mode C or S transponders or ADS-B Out equipment may be enough to satisfy the requirements. Where that may go with Advanced Air Mobility/Urban Air Mobility concepts is another issue, but that is being worked on by the developers and the regulators.

The third issue is more or less moot, as UTM or U-space as conceived are clearly a new kind of airspace, but one that is to be integrated into or compatible with legacy ATM systems. There are no plans to replace existing airspace categories or requirements with a UTM system. Doing so would be a nearly insurmountable task, requiring the overhaul of ICAO's Rules of the Air and virtually every other ATM system around the world.

Integration, not replacement, is the goal. A functional breakdown of the domains in the global air traffic management system could include:

- National airspace
- Low level airspace
- Commercial vs. recreational operations
- Conflict resolution between UAS and manned aircraft
- Detect and avoid or sense and avoid capability
- Autonomous operations
- BVLOS vs. VLOS
- Size, weight, speed, and intent (mission)
- Equipment requirements (RID, ADS-B, other)
- Coordination management with USS and DSS
- Requirements for UAS Service Suppliers (USS), Supplemental Data Service Suppliers (SDSP), Flight Information Management Systems (FIMS), Urban Air Mobility (UAM), and Advanced Air Mobility (AAM)
- Regulations
- Personnel
- Technical readiness levels of all new components
- Public perception and privacy

This chapter has proposed a simple framework (or more modestly a set of questions) for a methodology for assessing the risk inherent in the integration of remotely powered aircraft into national airspace systems. It is not the only method, nor is it as comprehensive as the many fine books and articles that have been written on the subject.

References

Craven, J. (2019). *The Meaning of "Form Follows Function."* Thought Co. website, available at: https://www.thoughtco.com/form-follows-function-177237. (accessed October 7, 2021).

FAA Order 8000.369C Safety Management System available at: https://www.faa.gov/documentLibrary/media/Order/Order_8000.369C.pdf

Vaughan, D. (2016). *The Challenger Launch Decision: Risky Technology, Culture and Deviance at NASA*. Edition. Chicago, USA: The University of Chicago Press.

8

Cybersecurity and Cyber Resilience

The World Economic Forum's "The Global Risks Report 2021," 16th edition, summarizes the results of its Global Risks Perception Survey (GRPS). "Among the highest likelihood risks of the next ten years are extreme weather, climate action failure and human-led environmental damage; as well as digital power concentration, digital inequality and cybersecurity failure. Among the highest impact risks of the next decade, infectious diseases are in the top spot, followed by climate action failure and other environmental risks; as well as weapons of mass destruction, livelihood crises, debt crises and IT infrastructure breakdown."

The survey results shown in Figure 1 of the report lists cybersecurity failure as the fourth highest risk in the "Clear and Present Dangers" category (short term risk, 0–2 years). In the survey's "Knock-on Effects" category (medium term risks, 3–5 years), cybersecurity ranks eighth on a list of ten risks. [Author's note: "Knock-on" is an interesting use here, the term having a specific meaning in the game of rugby union.]

Again, from the WEF report: "Public-key cryptography, or asymmetric cryptography, is a cryptographic system that uses pairs of keys. Each pair consists of a public key (which may be known to others) and a private key (which may not be known by anyone except the owner). The generation of such key pairs depends on cryptographic algorithms which are based on mathematical problems termed one-way functions. Effective security requires keeping the private key private; the public key can be openly distributed without compromising security."

So, the question is: What are "cybersecurity" and "cyber resilience?" And why is this topic important to our understanding of airspace integration, UTM, U-space, Advanced Air Mobility, Urban Air Mobility, and other elements of a global ATM system?

Simply stated, none of these air traffic management systems or concepts of operations for the development of these systems can exist without electronic communication or interconnectivity. A breach or breakdown of any of the critical elements of these systems can be catastrophic. At the very least, an interruption or complete failure of the system could have local, regional, or global impacts on commerce, safety, security, and the economy of a state or region.

The Global Risk Report also includes a report (represented by a graph on page 54) of the total number of significant cyberattacks from 2006 to 2020. Not unexpectedly, the US had the highest number (156), followed by the UK with 47, India with 23, Germany with 21, and South Korea with 18. It should come as no surprise that China and Iran only had 15, and Russia, with 8, and North Korea, with 5, were near the bottom of the list, as those four nations are the source of most of the cyberattacks on other countries. A study by the Center for Strategic and International Studies (CSIS), a think tank based in Washington, DC, stated that Russia and China were the source of the vast majority of cyberattacks on other countries. A 2018 report from CrowdStrike[*] labeled China as the biggest state sponsor of cyberattacks on the West.

It should also be noted that all of those countries are ICAO Member States, subject to ICAO's rules and standards, including cybersecurity. Their ICAO membership compels their compliance with ICAO's rules or face expulsion from the organization. Yet they are still Member States and they still either initiate or tolerate cyber attacks on other nations or organizations.

In December of 2020 and May of 2021 Sandia National Laboratories (a multimission laboratory managed and operated by National Technology and Engineering Solutions of Sandia, LLC, a wholly owned subsidiary of Honeywell International, Inc., for the US Department of Energy's National Nuclear Security Administration under contract DE-NA-0003525) hosted a series of virtual cybersecurity events called "Meetings of the Minds," to coordinate the "best thinking on how the nation should respond to changing threats." These two events featured expert presenters from the National Risk Management Center at the Cybersecurity and Infrastructure Security Agency, the Cyberspace Solarium Commission, and other groups on the front lines of cybersecurity to discuss a national policy to confront the threats faced by governments and private enterprises.

> "In the May session, following several high-profile national cyberattacks, Sandia brought together panelists from CISA, Google, Microsoft and other institutions to talk about emergent research and development. Sandia systems analyst Eva Uribe highlighted the complex nature of cyber conflict and competition."
>
> "Severity of attack has been used as a guiding principle for how to respond. We are trying to protect against a Pearl Harbor," she said. "But an overwhelming number of smaller attacks can potentially target the things that are important to us – power, transportation, wireless networks and, as we saw, the Colonial Pipeline."
>
> "In May, the operator of the Colonial Pipeline paid $4.4 million to hackers who successfully attacked the system that transports gasoline from the south to the northeast corridor. Meeting participants in December learned that 85% of the nation's water, electricity and natural gas infrastructure are managed by local and regional agencies that lack budgets to pay for high-end cybersecurity. Data shows that ransomware attacks increased by 41% in 2019, with successful penetrations in more than 205,000 business and local entities and attacks are still coming."
>
> David White, director of Sandia's Information Operations Center, emphasized the need for preparedness, starting with these kinds of meetings to coordinate the best and brightest in cybersecurity.
>
> "I don't think we can ever fully deter our adversaries. Resilience is really the core interest of the nation," David said, adding that Sandia is already helping confront these threats. "It's an insidious problem. I think it requires a national response. Sandia can certainly advise about new threats and opportunities, but the best role for our brilliant people is solving the hardest technological problems."

Cryptography relies heavily on random numbers to generate unbreakable keys. Research is underway at the Sandia Labs to explore better and less expensive ways to generate random numbers to make encrypted communications even more resilient than they are with current technology.

Security of the air traffic management system is a fundamental component of the overall safety of the system. Aviation is a safety-critical business. Airspace integration efforts contemplated by UTM, U-space, UAM, AAM, and other concepts of operations intended to mix manned and unmanned or autonomous aircraft in unsegregated airspace cannot succeed without data and system security. The evolution and expansion of digital aviation infrastructures and services have played significant roles in increasing capacities and efficiently managing global airspace. However, this growth has come with a downside with regard to interoperability and cybersecurity and resilience. The emergence of new technologies (unmanned aircraft in particular) that have significantly different operational characteristics and capabilities has made the air navigation system more complex. The creation and implementation of new airspace environments has increased complexity by an order of magnitude, and along with this a more dynamic and vulnerable system of communication in a critical infrastructure has emerged.

Describing the Threat

Cybersecurity is a top priority for UTM or U-space developers. In this context, security refers to the protection against threats from both intentional and unintentional acts that threaten people and/or property in the air or on the ground. Security risk management goals include balancing the needs of the members of the UTM or U-space communities who desire access to the airspace with the need to protect stakeholder interests and assets, including the civil aviation authorities, public safety entities, airspace participants, and the general public. Threats to aircraft or ATM systems can be mitigated by information provided by UTM capabilities to aviation and security authorities. Protecting data exchanges among participants in the UTM system from corruption or denials of service is the core mission of robust cybersecurity strategies.

A key element of cybersecurity is the integrity of the information being exchanged between participants in the system. This component is the focus of an ICAO study group that is tasked with developing a standard for ensuring information exchange integrity across the whole of the aviation world in a structure that is uniform and consistent for all users, participants, and stakeholders. This ICAO effort will be discussed in a later section of this chapter.

The Honorable Ted Lieu, representing California's 33rd Congressional District, stated in a recent interview that the US is "way behind" in cybersecurity, due to many weak links in contractors' supply chains. Referring to the "Internet of Things" (IoT), he stated that cyber certification in the private sector has created significant vulnerabilities that are being addressed by pending bipartisan legislation (a significant development in a sharply divided nation). Representative Lieu advocated for the idea that "space" should be considered a "critical infrastructure," with the same protections as agriculture and transportation, to name just two others that fall into that category. Since GPS, GNSS, and other internet-based forms of communication all rely on satellites and other airborne systems, "space" (as in the zone above and around our planet where there is no air to breathe or to scatter light) should be elevated to the highest level of cybersecurity.

In that same broadcast, Christopher Krebs, founder and CEO of the Krebs Stamos Group, discussed "attack surfaces," or literally any device that is connected to other devices or users via electronic transmissions, both wireless and connected by cable. In defending ransomware attacks in particular he recommends an overall strategy of: (1) assuming that there will be a breach of a system and (2) developing a layered approach to the defense of the system. Mr. Krebs used an interesting term: "Blast Radius." In other words, unauthorized entry into an otherwise protected system, which in turn is probably connected to other systems, can have a wide-ranging effect on a much larger attack field than the original target. As will be discussed elsewhere in this chapter, a simple example would be an individual user at a private company or a government agency allowing their login information to be compromised (by using a proprietary website to access an unauthorized website or allowing another individual to use the person's unique login credentials, etc.). The immediate effect would be to give the attacker broad access to many other data sources beyond the entry point, thus compromising multiple proprietary (or encrypted) systems that are otherwise protected from hostile intrusions. The preferred strategy is to erect barriers to unauthorized entry (through a wide variety of mechanisms such as multifactor identification, training, etc.), and then to implement resilience mechanisms, allowing for effective recovery capabilities.

The 2018 ICAO meeting in Mexico featured a presentation by The Thales Group entitled "Cyber-Security for Air Traffic Management." In discussing the proliferation of cyber attacks on aviation assets, the presenter identified six examples, presumably drawn from a significantly longer list:

- Ransomware attacks blank out screens at UK airport
- Hackers deface airport screens in Iran with anti-government messages
- FBI warns of cyber-thieves targeting aviation
- Cyber-chaos at Heathrow
- Access to an airport's security system sold on the dark web
- Ransomware targets Civil Aviation Authorities
- And so on ...

Deconstructing this horror show list of attack vectors down even further, the presentation identified several areas of anticipated (or existing) vulnerabilities:

- Insertion of an infected USB key on an Online ATC System
- Login usurpation on an ATC system technical position
- Remote intrusion in an ATC LAN (local area network)
- Spoofing of an ADS-B radio signal
- Spoofing of radar data over an interconnection network
- Controller or pilot usurpation over Datalink
- Malware injection in maintenance
- Spoofing of GPS time
- Denial of service on the Aeronautical Information Service
- Malware injection in the supply chain

These two lists of past and possible future cyber attacks are just examples of the countless scenarios that challenge the experts and developers of cybersecurity systems. How the global and regional aviation authorities and policy makers are responding to these threats is the focus of the following sections.

To cover every effort around the world would require an entire book of its own. The following discussion is intended to highlight some of the more significant efforts undertaken by international organizations and governments. Further reading listed at the end of this chapter may be useful to fill in any gaps.

ICAO

As discussed above, cyber threats are a growing global concern to all forms of electronic information sharing, and the aviation infrastructure is near the top of the list of the most vulnerable systems. It is critical to aviation safety that the integrated and highly automated air traffic management system currently in place is capable of providing trusted information exchanges throughout the global system. This requirement calls for a framework where all parties to a communication are able to confidently identify themselves to one another and that the information exchanged cannot be altered by unauthorized parties.

ICAO's Secretariat sought to develop a globally harmonized and interoperable international aviation trust framework for the secure and efficient exchange of digitally connected information. To advance that goal, ICAO established the Trust Framework Study Group (TFSG) in 2019 to develop the desired framework through three working groups: operations, digital identity, and network. These working groups focused on the development of policy and guidance material to enable trusted ground-to-ground, air-to-ground, and air-to-air exchanges of information among all current and prospective aviation stakeholders. These working groups have reviewed the concept of operations, defined use cases, developed a digital certificate policy, identified the security and access control requirements for a global resilient aviation interoperable network, and have presented a series of working papers (WPs) to the Council and the Assembly (more on those efforts are given below).

Another related ICAO effort was the establishment of an ICAO study group, which is tasked with ensuring uniform information integrity across all aspects of aviation. Multiple states, industry, and aviation stakeholders have collaborated to define a cybersecurity network and create policies for the International Aviation Trust Framework (IATF).

IATF

IATF's task is to create an international operational network and identity policy framework, creating a Global Resilient Aviation Information Network (GRAIN). GRAIN is a network of networks interconnecting aviation stakeholders for all information exchanges. A chart on the referenced document ("Global Resilient Aviation Network Concept of Operations"), Figure 6 in the document, "... depicts the cybersecurity and network policy relationship with reference to UTM stakeholders. Not all networks that operate under the IATF network policies are necessarily interconnected. Some network connections use the

IATF network policies without being 'connected' to GRAIN; other network connections use the identity policies without the network policies." (It should be noted that this document is available in draft on the ICAO website, was last modified on 21 November 2019, and evidently is subject to amendments before the final document is released.)

> "All UTM stakeholders using IATF policies use an IATF compliant Registration Authority (RA) to perform the vetting and proofing of the identities. In addition, all UTM stakeholders using the IATF policies use an IATF compliant Certificate Authority (CA). The RA and CA can be implemented by commercial entities. Identities issued by different CAs under IATF policies are interoperable and can trust each other. The trust relations between identities can be managed by the individual stakeholders and by application domains."

The document is comprehensive and instructive for anyone interested in more in-depth understanding of the issues surrounding cybersecurity and resilience as they impact civil aviation, and particularly the challenges to the development of UTM or U-space protocols. The framework set forth in this document offers an excellent overview of the cybersecurity issues confronting the global aviation community, and warrants summarizing here.

The subtitle of the document is "For a secure and trusted exchange of information." After outlining the current and future information exchange environments, as well as the need for a global resilient aviation network, the drafters go on to outline the "global resilient aviation network operational concept."

The architecture of the CONOPs includes the network, the systems that define the network, and the applications that will run on the system.

The network is shown in a clever "poodle diagram," which describes the cyber resilience perspective of an aviation network (the diagram roughly resembles a poodle dog). Messages will pass through service exchanges, which will have multiple cybersecurity controls and processes. Among the services performed by the exchanges will be "sanity checks" to support policy enforcement protocols (such as verification and correlation, message context specific attributes, destination, time of day, etc.).

Systems will be configured to only permit limited access to network ports. Access will be granted only through properly authenticated, privileged, and secured sources, thus eliminating cyber attacks through malware and similar sources.

Applications will run on systems and will communicate with one another. Mutual authentication (through using a digital identity) must be performed before two applications can initiate a communication session. Communications between applications will be facilitated by message exchange, through message integrity signatures (digital identities). If any of the required elements are missing or incorrect, the message will be dropped. Message exchange protocols are provided by Java Message Service (JMS), Advanced Message Queuing Protocol (AMQP), which do not support a message integrity signature, or by virtual private network (VPN) or transport layer security (TLS), which do.

Cyber resilience consists of six (6) core elements: identification, protection, detection, mitigation, recovery, and compliance.

The first step in the identification phase is for all stakeholders to ascertain the elements of their systems that support critical functions for safety of flight and their known cyber vulnerabilities and associated risks. This process is critical for certification authorities (CAs) in managing digital identities, network service providers, and communications service providers. These three provider categories play a crucial role in the aviation system and can create vulnerabilities derived from identification lapses. The cyber vulnerability information exchanged between aviation stakeholders is essential to support cyber resilience in the aviation network, as the information can be used to strengthen cyber defenses and bolster resilience and recovery.

"Protection" from potential cyber events should also include authentication or access control, data security, information protection processes and procedures, maintenance of the systems, and multiple layers of protective technology.

"Authentication" and access control will be supported by multiple communication levels, which should meet minimum performance requirements based either on industry best practices or standards and regulations, where applicable. This again will rely upon mutual authentication strategies, as described above. Maintenance of large amounts of digital identities that enable an entity to communicate globally will be a major challenge to stakeholders.

"Data security" includes both integrity (unauthorized modification of communication) or confidentiality (protection from unauthorized data inspection). Mutual authentication with digital identities will enable the parties to use their identity to sign or encrypt data as needed.

"Information protection" processes and procedures contemplate stakeholder documentation of data security controls of the procedures used to protect critical aviation services. Global recommended practices and procedures will define the criteria to meet those requirements.

"Maintenance" includes timely updates of applications, software, firmware, and security configurations, based upon previously identified cyber vulnerabilities.

"Protective technology" includes policy enforcement points, firewalls, intrusion protection systems, network flow controls, and other measures that match the network architecture and cyber risk assessments.

"Detection" really means enabling the timely discovery of cybersecurity threats or attacks. Effective detection of intrusion into the system is enhanced by each stakeholder's participation in detecting and reporting events so that all the relevant information can be correlated across the network. This cooperation in turn should facilitate the isolation and detection of security anomalies.

Two of the tools to be employed in this regard are Certificate Authorities (CAs) and Bridge Certificate Authorities (BCAs). CAs detect anomalous discovery and validation requests. BCAs detect anomalous certificate policy mapping and cross-certificate validation requests, by mapping origin, destination frequency, policy, time of day, and other indicators that do not satisfy the certificate policy requirements. These applications also serve to detect anomalous authentication and access requests and message integrity events. Air-to-ground and ground-to-ground service providers will detect irregular network flow, denial of service, and intrusion attempts. Correlation of these events is critical to detecting the full scope of a cyber attack.

Simply stated, "mitigation" or response refers to the ability of the network to contain the effects of a cybersecurity event. Stakeholders need to document the appropriate measures taken to detect the event and to communicate mitigation processes to other essential parties. Documentation should include the various technical processes undertaken to isolate, analyze, and block cybersecurity attacks. Based upon receipt of evidence of involvement in a cybersecurity incident, a CA may revoke one or more identities. Similarly, a BCA may revoke the cross-certification of all identities of a CA if it detects or has been notified that the CA has been compromised. A BCA may also place limits upon certificate policy mapping for certificates from a policy domain if there are issues of trustworthiness of the domain. In addition, a ground-to-ground or air-to-ground service provider may block network flow generated by a questionable source to protect the network from a Distributed Denial of Service (DDoS) or impersonation attack.

"Recovery" refers to the process of restoring capabilities or services that are impaired as a result of a cyber event. This also requires stakeholders to document recovery activities and to share those processes with other users.

The document goes on to describe the enablers of a resilient network. They are: information security management systems, a public key infrastructure, iPv6 addressing, and a Domain Naming System.

An Information Security Management System (ISMS) envisions a resilient operating network infrastructure design that requires proper operation to ensure protection. The process of implementing cybersecurity related systems and timely reassessment of risks is supported by a set of existing standards (ISO 27000–27005). The standards require monitoring of cybersecurity related events and discovered vulnerabilities, which will be shared among stakeholders connected to the global resilient operating network.

The "public key infrastructure" (PKI) refers to a set of rules, policies, and procedures that are required "to create, manage, distribute, use, store, and revoke digital certificates and manage public-key encryption." PKIs enable the secure electronic transfer of digital information. This technology is necessary where a simple password is an inadequate method of authentication, and a more robust method is required to confirm the identity of parties to a communication and to validate the authenticity of the information being transferred.

PKIs bind public keys to specific identities of people and organizations, and is an element of cryptography. Binding is accomplished through registration and issuance of certificates by a CA. The function may be performed by automation or manually under human supervision, depending upon the assurance level of the binding. "Registration authority" (RA) derives from a PKI's validation of a correct registration. The RA's role is to accept requests for digital certificates and to verify the identity of the entity making the request.

To ensure the integrity of the identification, each entity must be uniquely identifiable within each CA domain, as determined by information about that entity.

A PKI can also serve as a system for the creation, storage, and distribution of digital certificates (which in turn verify that a particular public key belongs to a particular entity). The digital certificates created by PKIs map public keys to entities and securely store the certificates in a central repository. PKIs can also revoke the certificates if necessary.

The components of a PKI are:

- A certificate authority (CA) that stores, issues, and signs the digital certificates
- A registration authority (RA) that verifies the identity of entities requesting that their digital certificates be stored at the CA
- A central directory (a secure location in which to store and index keys)

- A certificate management system managing access to stored certificates or the delivery of the certificates to be issued
- A certificate policy describing the PKI's requirements concerning its procedures, thus allowing outsiders to analyze the PKI's trustworthiness

As indicated above, the certificate authority digitally signs and publishes the public key that is bound to a specific entity. The CA's own private key accomplishes this task in a manner that other entities' keys rely on trust in the validity of the CA's key. The key-to-entity binding is established by software or under human supervision, depending on the level of assurance the binding enjoys. The CA regularly publishes a revocation list of the keys that are no longer valid and also hosts an online Certification Status Protocol (OCSP) responder (which is used by relying parties to obtain revocation status of a public key). These functions can be delegated to sub-CAs, in which case the main key is designated the root CA. The root CA then signs and publishes the key generated by the sub-CA.

An aviation PKI can be created by any one of three strategies. A single root CA, with sub-CAs for every state, region, or organization requires existing aviation root CAs to become sub-CAs and to reconfigure every system employing the CA to adjust to the new PKI hierarchy. The PKI is then legally responsible for one root CA rather than a collection of sub-CAs.

A second option is a root CA per State, region, or organization, supported by individual bi-lateral agreements to allow for cross-certification between each root CA. This structure requires the establishment and maintenance of any number of root CAs in the PKI. The discovery and validation of multiple information paths is time consuming and prone to error, and may not be favored by large organizations.

The third alternative is the Bridge Certificate Authority (BCA), which the authors state is the most scalable approach to unify existing and future PKIs. A Bridge Certificate Authority is a third-party service provider that facilitates harmonization across the aviation community on standards, processes, and governance for a trusted entity. In this structure each root CA continues to sign and publish its own keys and manage its own liabilities. "Although the validation of the keys is still complex, there is only one deterministic path to validate." A BCA requires a trust framework for cybersecurity specific standards and policies, supported by an enforceable governance regime, all intended to facilitate a globally interconnected community participating in a common operating environment.

A digital certificate or identity (aka a public key certificate) describes an electronic document that is intended to prove the ownership of a public key. The certificate includes information about the key, the identity of the owner, and the digital signature of the entity that verified the certificate's contents. If valid, and the software tool that examined the certificate trusts the entity that verified the certificate's contents, the key can be used to securely communicate with the certificate owner. Thus, a process of a public key certificate validating another content validator's own validation of a digital signature on a certificate is at least a two-level security check.

This multilayered common-identity validation process enables airspace managers to confidently identify all airspace users with a unique identification assigned to every aircraft (and potentially many of its component parts). An aircraft can be accurately tracked through all phases of flight using digital identifiers that consider the aircraft itself and the context in which it is operating. This is a vast improvement over identifying aircraft by tail numbers, flight numbers, or pilot-to-ATC communication.

An advance in technology in Internet Protocol (IP) address identification will enhance the network's capability for identifying a communicating party's individual network interface, locating the party's position on the global network, and thus facilitating routing of IP messages between communicating parties. IP addresses contain information on origin and destination of a message, such as bank routing numbers that allow for digital transfers of information.

The previous wide-use IP address standard was iPv4. A new standard is now available to the aviation world. Internet Protocol version 6 (IPv6) is the most recent version of the Internet Protocol (IP), the communications protocol that provides an identification and location system for computers on networks and routes traffic across the internet. IPv6 was developed by the Internet Engineering Task Force (IETF) to deal with the long-anticipated problem of IPv4 address exhaustion.

The new version resolves incompatibility issues inherent in iPv4 and has double the bit address space that was available in iPv4. The two IP versions are mostly incompatible, but can exist on the same network. An iPv4 address can only communicate with another iPv4 address, and the same goes for iPv6.

Another digital identification or identity verification method is the Domain Name System (DNS) and generic top-level domain (gTLD). DNS is one of the underlying services used in IP networking infrastructures. It bridges the gap between human readable addresses and technical alphanumeric addresses, which are gibberish and difficult to memorize. DNS Name Servers contain information about a small portion of the domain name space. The domain name information provided by DNS is meant to be available to any computer located anywhere on the internet. DNS supports security goals related to data integrity and source identification. It is an essential service, but vulnerable to hacking or intrusion and redirection of communication.

To combat hostile actions towards network services, DNS uses an intrinsically protected service called Domain Name System Security Extensions (DNSSEC), which is a collection of Internet Engineering Task Force (IETF) specifications for securing information derived from the DNS as used on IP networks. Another choice is to use a private DNS service that is not accessible outside of the aviation ecosystem.

A generic Top-Level Domain (TLD), a top-level domain maintained by the Internet Assigned Numbers Authority (IANA), is also required. A top-level domain can be thought of as the last security level for every fully qualified domain name.

Global Resilient Aviation Network Concept of Operations – Trust Framework

> "Establishing trusted communications links between the aviation stakeholders and being able to trust the exchanged information are fundamental to the evolution of the aviation system and to manage the complexity of the future demand while keeping the defined levels of safety."

The creators of this document stress that this concept of trust relies on two fundamental elements: the identity of the communication parties and the integrity of the message contained in the information exchanged. They recognize that the aviation ecosystem is undergoing a digital transformation, which brings with it significant challenges with respect to cyber security. Although not mentioned in the document, the new concepts surrounding UTM/U-space, urban and advanced air mobility, and the rapidly evolving state of aviation technology, have added additional layers of complexity to the global aviation network.

The authors propose three principles underlying the global trust framework:

- Interoperability assurance through a common trust framework, as provided by seamless and efficient messaging service architecture operating on a global scale.
- Maximized resiliency of communication by using design concepts that limit the threat surface (a term that others have defined as any and all devices that can or do connect to the Internet – in other words, billions of potential entry points). This requires compartmentalization of the network and the applications, layering protections in the infrastructure and placing limits on the number of systems participating in the network.
- Robust identification, authentication, authorization integrity, and confidentiality.

It goes without saying that cryptography principles and practices are necessary to establish technical trust among parties and stakeholders in the global trust framework.

The concept of network "cyber hygiene" relies upon implementation and abiding by an agreed upon cybersecurity policy, as found in the aforementioned ISO/IEC 27000 series of standards for information security management. Thus, networks and their operators must be ISO 27000 certified to demonstrate compliance with cybersecurity policy. Regular audits performed by ISO 27000 certified accreditation bodies are necessary, and the certification reports are reviewed by ICAO and provided to all network operators. Trusted digital identities are achieved by adhering to administrative requirements, legal agreements, and operational functions as defined by the organizations established to serve those needs. A graphic roadmap for accomplishing this is found at Figure 8 in the document.

Governance of the framework outlined above involves management of three sets of agreed upon documents:

- *Legal agreements* define the risks and liabilities as agreed between the members.
- *Certificate policies* define the procedural and operational protocols that the trust framework requires member organizations to follow when registering, using, managing, verifying, validating, and using digital identities.
- *Technical specifications* specify the tools, standards, and interfaces used to operate the framework of a trusted identity.

The governing body performs audits and monitors, reviews, and reports on compliance with the certificate policy.

On the operational side, sets of administrative documents and operational functions describe the rules of the road, so to speak, to ensure conformance with the network requirements. Administrative documents consist of registration policies that document the processes for identification of physical identities to conform registration of a digital identity with the physical identity. Certificate practice statements document the processes used by credential providers to issue, manage, and validate digital identities. Service agreements document functional and performance requirements for services and technical interfaces between service providers.

Operational functions include accreditation and certification processes to evaluate, describe, test, and validate digital entities; bridge and credential exchange to establish trust between digital identities issued by different credential service providers; and attribute exchanges perform real-time exchange of identity attributes between attribute, credential, and identity service providers.

Private iPv6 "address blocks" provide many advantages for the aviation community. They are an effective way of managing integration issues such as those posed by unmanned aircraft. A private IP address cannot by routed on the Internet. A globally unique aviation address range allows private ISPs to offer an additional layer of isolation by rejecting direct routes to that private block.

The advantages derived from the use of private IP address blocks include:

- Allowing connectivity and communication across private and public infrastructures
- Allowing organizations to use private IP addresses for internal operations
- Allowing organizations to use their own registered iPv6 address for public and private use across the aviation infrastructure
- Standardizing methods to limit risk to IP infrastructure by limiting the range of IP addresses
- Allowing for future connectivity in a future communications infrastructure
- Simplifying protection of network assets

Returning to DNS, the authors state: "A private DNS for the aviation generic top-level domain (gTLD) is necessary for IP to be effectively utilized across the global aviation community and to limit and isolate the cyber risk that can come with the use of DNS. A DNS structure that follows the IP addressing schema and meets the needs of the aviation community needs to be defined."

The operation of the network will be provided by a consortium of traditional aviation network providers, including new entrants providing LTE-based services or low earth orbit networks supported by balloons or drones. The consortium will comply with global standards for network service requirements, network interface requirements and policy, and network security requirements and policy. This will be facilitated by global multilateral agreements for connecting networks.

Network service requirements are subject to the following parameters:

- Available bandwidth
- Maximum latency based on message size
- Availability, mean time between failure and mean time to restore over a defined period of time
- Network protocol support
- Local scope or global scope

"Network interface requirements and policy define how the network providers within the consortium interconnect their respective networks. The requirements specify the technical interface standards for the interconnection. The network interface requirements will specify the use of global IPv6 addresses for aviation between the network providers and the network users. The policy describes the process to follow for interconnecting providers' networks, the roles and responsibilities in the process, and the legal liabilities for each network provider. The policy also describes how to obtain and assign IPv6 addresses to the network."

Network security requirements and policies define standards for security control network service providers that users must follow and the process that is required to be used to prove compliance.

"Summary of Impact:

The full connectivity enabled by digital transformation of the aviation system is occurring now, and will continue. The question is whether this proceeds in a coordinated fashion that enhances interoperability and reduces the threat surface or not. Communication service providers, aircraft manufactures, and avionics producers, are all putting in place their own systems of identity and trust as a matter of necessity. That means in the near future, an aircraft may need different digital certificates to communicate with its satellite communications provider, retrieve data from the airline operations centre, update its avionics, download engines monitoring data and other functions. The potential number of proprietary secure links is nearly endless. This patchwork of disparate efforts to reduce the threat surface to air and ground operations will add complexity to the system that will be costly to maintain and will offer a myriad of gaps for adversaries to exploit.

Similar problems are already being encountered on the ground as ANSPs, Airports, and other service providers attempt to exchange information as outlined in the Global and Regional Air Navigation Plan (GANP/ANPs). States and Regions are putting in place their own internal systems of identity and to allow them to operate. These systems will not be able to connect to internal or external entities to the aviation community unless trusted mechanisms for digital identification are recognized and put in place at a global level.

It is also important to note the opportunity that is about to be lost regarding the new entrants into the system. Across the world, several Civil Aviation Authorities are responding to the massive influx of UAS. Many are putting in place registries systems and there are ongoing debates around the possibility of an electronic identification system. In the absence of direction global direction, different manufactures and different States will take different approaches. A globally acceptable system of identity and trust would channel the innovation that drives this emerging industry in the direction of interoperability, and increase levels of safety in a connected environment.

There are many aspects of the concept of a trust framework that appear daunting, and it will take years for the full operational vision to be realized. However, substantial benefits will be realized long before the system becomes fully operational. At this moment, the entire aviation community is responding to the increase in the threat surface with different and uncoordinated actions. This divergence, and its effect on interoperability, is crucial for the aviation system. This divergence will begin to reverse once a global approach is agreed among the aviation community."

ICAO Council and Assembly Documents

Assembly – 40th Session, Executive Committee 2/8/19 Working Paper A40-WP/427

Proposal for ICAO Governance of Cybersecurity and Resiliency (presented by the United States)

Executive Summary

"Cybersecurity and resiliency in the aviation ecosystem is a multidisciplinary issue that affects or will affect nearly every aspect of global aviation. Due to the complexities and the reliance on shared digital information and communication, the need for cybersecurity and resiliency becomes more vital with every advancement in technology and the continuous modernization of the aviation ecosystem.

While ICAO, Member States and industry are diligently working to address cybersecurity and resiliency issues, the current approach at ICAO lacks appropriate governance and focuses on individual sectors and varying expertise instead of reflecting the holistic global aviation ecosystem.

To address these issues, the United States recommends that ICAO establish a Council Technical Committee on Cybersecurity and Resiliency to centralize governance and properly address cybersecurity and resiliency in a holistic approach. The Committee will manage policy and integration of industry standards while evaluating potential development of technical Standards and Recommended Practices."

"Action: The Assembly is invited to:

a) Request that the ICAO Council establish a new Council Technical Committee on Cybersecurity and Resiliency as proposed in this paper;

b) Urge States to support the proposed ICAO Cybersecurity Strategy developed by the Secretariat Study Group on Cybersecurity; and

c) Urge States to support the work of the Trust Framework Study Group."

This proposal is the United States' suggestion to enhance governance of cybersecurity and resilience by taking a holistic approach to engaging the larger community in managing policy and integration of industry standards. It generally reflects the desires of the majority of the Member States, if not all of them.

Declaration on Cybersecurity in Civil Aviation (Dubai, 2017)

"We the Officials and representatives from the States and regional and international organizations listed in the Annex,

Declare that:

1. It is the responsibility of States to act in such a way as to mitigate the risk posed by cyber threats, to build their capability and capacity to address such threats in civil aviation, and to ensure their legislative framework is appropriately established to take action against actors of cyber-attacks;
2. Cyber capabilities applied to aviation should be used exclusively for peaceful purposes and only for the benefit of improving safety, efficiency and security;
3. Collaboration and exchange between States and other stakeholders is the sine qua non for the development of an effective and coordinated global framework to address the challenges of cybersecurity in civil aviation;
4. Cybersecurity matters must be fully considered and coordinated across all relevant disciplines within State aviation authorities;
5. Cyber-attacks against civil aviation must be considered an offense against the principles and arrangement for the safe and orderly development of the international civil aviation; and
6. The ratification and entry into force of the Beijing Instruments would ensure that a cyber- attack on international civil aviation is considered an offence, would serve as an important deterrent against activities that compromise aviation safety by exploiting cyber vulnerabilities, and therefore it is imperative that all States and ICAO work to ensure the early entry into force and universal adoption of the Beijing Instruments, as called for in ICAO Assembly Resolution A39-10: Promotion of the Beijing Convention and Beijing Protocol of 2010; and

Reiterate our commitment to the development of a robust, efficient and sustainable civil aviation system."

ICAO "Update on the Progress in Cybersecurity," Twenty-Second Meeting of the AFI Planning and Implementation Regional Group (APIRG/22) (Accra, Ghana, 29 July–2 August 2019)

This information paper presents an update on the ongoing cybersecurity work of ICAO and provides information related to a Cybersecurity Strategy Draft Assembly Resolution, as well as appropriate cybersecurity awareness and training.

The report noted that the 39th Session of the ICAO Assembly reaffirmed the importance and urgency of protecting civil aviation's critical infrastructure systems and data against cyber-attacks and obtain global commitment for action by ICAO, its Member States, and industry stakeholders, with a view to collaboratively and systematically addressing cybersecurity in civil aviation and mitigating the associated threats and risks. Resolution A39-19, Addressing Cybersecurity in Civil Aviation, identified the actions to be undertaken by States and other stakeholders in this regard. The 39th Session of the ICAO Assembly also instructed ICAO to develop a comprehensive cybersecurity work plan and governance structure.

To meet these objectives, ICAO established the Secretariat Study Group on Cybersecurity (SSGC) under the lead of the Deputy Director, Aviation Security and Facilitation (DD/ ASF). The SSGC is monitored by the Secretariat Senior Management Group on common safety and security issues, chaired by the Secretary General of ICAO.

The SSGC focuses its attention on cybersecurity strategy. The strategy aims for: (a) the protection of civil aviation and the travelling public from cybersecurity threats that might affect the safety, security, and trust of the air transport system; (b) maintaining or improving the safety and security of the aviation system in preserving the continuity of air transport services; (c) States to recognize their obligations under the Convention on International Civil Aviation (Chicago Convention) to ensure the safety, security, and continuity of civil aviation, taking into account cybersecurity threats; and (d) coordination of cybersecurity measures among State authorities to ensure effective and efficient management of cybersecurity risks.

Among other relevant topics, cybersecurity awareness and training continue to be of great interest and concern at the global and regional level. Various regional awareness and training events in the form of workshops, seminars, and training courses have started to evolve through initiatives by States and the ICAO Regional Offices.

ICAO Council 224th Session, Fourth Meeting C-DEC 224/4, "Mechanism to Address Cybersecurity in ICAO"

This open virtual meeting was held on 27 October 2021. The committee approved the proposed Terms of Reference for the Cybersecurity Committee, as contained in Appendix A of C-WP/15255, and further agreed to rename the Cybersecurity Committee as the Ad hoc Cybersecurity Coordination Committee. Among other topics, the committee took note of the Terms of Reference for the Cybersecurity Panel and affirmed that the resources necessary to support a new cybersecurity governance structure should be allocated as a "matter of high priority." The committee also discussed updates on ICAO cybersecurity activities, and requested the Secretariat to update the status of activities under the Cybersecurity Action Plan (which is not publicly available, only to Member States' verified representatives).

The importance of this Summary of Decisions is to confirm that ICAO assigns the highest priority to cybersecurity and resilience issues, mirroring what appears to be more or less a global consensus.

ICAO Thirteenth Air Navigation Conference, Montréal, Canada, 9-19 October 2018, AN-Conf/13-WP/27, "Cyber Resilience"

This working paper's Executive Summary states:

> "The importance of an appropriate approach to cyber resilience in civil aviation has been recognized at the global level and is a crucial matter to be dealt with in the 2019 edition of the Global Air Navigation Plan (GANP) (Doc 9750). Successfully managing cyber resilience in an increasingly interconnected aviation system requires a globally harmonized approach between all stakeholders. Without a globally coordinated trust framework to secure ground–ground, air–ground, and air–air exchange of information and commitment to building a culture of trust leading to the sharing of cyber-

related incident information, cyber resilience may not be assured. It is also recognized that despite the increasing level of reliance on information technology, human involvement is simultaneously a leading cause of, and the first line of defence against, cyber threats and for this reason operational staff should be trained to recognize general aspects of cyber threats and to report any suspected cyber threat or incident."

The last sentence highlights perhaps the greatest threat to aviation cybersecurity, and that is people. Poor hiring practices, inadequate training, and rogue actors within the system are the intangible factors that can defeat the most robust cyber identification and validation mechanisms, and may even impede the discovery and recovery process, as noted previously in this chapter. Resilience is only as good as the people who design and manage the network. The task of the developers is to anticipate those "blast surfaces" and design cybersecurity systems that cannot be corrupted by one single point of attack or vulnerability.

ICAO CANSO CANSO – Airbus Air Traffic Management Cybersecurity Policy Template Iintroduction, NACC/WG/6 – WP/23 August 2021

Executive Summary

"Cyber-attacks are a growing threat worldwide because of increased digitalization and the interconnectivity of systems. Civil aviation is particularly sensitive to this emerging threat due to its widely interconnected systems. Any disruption of systems due to a cyber-attack can seriously affect the safety and security of flights and the reputation of civil aviation in the public eye. As such, ICAO addressed this emerging threat to civil aviation through ICAO resolution A40-10 Addressing Cybersecurity in Civil Aviation."

Under the heading "suggested actions":

"The Meeting is invited to:
   a) take note of the information provided in the working paper;
   b) encourage States and ANSPs to develop their cybersecurity and strategic plan to ensure continued mission operations regardless of the cyber threat; and
   c) participate in cybersecurity assessment and evaluation."

CANSO released its "CANSO Cyber Security Risk Assessment Guide" in 2014. It is a comprehensive 48-page summary of the challenges to the aviation community presented by cyber threats and risks. Appendix A covers international standards (specifically ISO 27000 series, ISO 27005 Information security risk management (ISRM), and the NIST cybersecurity framework). Appendix B outlines a Risk Assessment Methodology. The document also provides a useful list of sources. (See References at the end of this chapter.)

ICAO Assembly Document A40-10

This Assembly resolution was released in 2019 and "*Urges* Member States and ICAO to promote the universal adoption and implementation of the *Convention on the Suppression of Unlawful Acts Relating to International Civil Aviation* (Beijing Convention) and *Protocol Supplementary to the Convention for the Suppression of Unlawful Seizure of Aircraft* (Beijing Protocol) as a means for dealing with cyberattacks against civil aviation."

The Assembly "*Calls upon* States and industry stakeholders to take the following actions to counter cyber threats to civil aviation:

a) Implement the Cybersecurity Strategy;

b) Identify the threats and risks from possible cyber incidents on civil aviation operations and critical systems, and the serious consequences that can arise from such incidents;

c) Define the responsibilities of national agencies and industry stakeholders with regard to cybersecurity in civil aviation;

d) Encourage the development of a common understanding among Member States of cyber threats and risks, and of common criteria to determine the criticality of the assets and systems that need to be protected;

e) Encourage government/industry coordination with regard to aviation cybersecurity strategies, policies, and plans, as well as sharing of information to help identify critical vulnerabilities that need to be addressed;

f) Develop and participate in government/industry partnerships and mechanisms, nationally and internationally, for the systematic sharing of information on cyber threats, incidents, trends and mitigation efforts;

g) Based on a common understanding of cyber threats and risks, adopt a flexible, risk-based approach to protecting critical aviation systems through the implementation of cybersecurity management systems;

h) Encourage a robust all-round cybersecurity culture within national agencies and across the aviation sector;

i) Promote the development and implementation of international standards, strategies and best practices on the protection of critical information and communications technology systems used for civil aviation purposes from interference that may jeopardize the safety of civil aviation;

j) Establish policies and allocate resources when needed to ensure that, for critical aviation systems: system architectures are secure by design; systems are resilient; methods for data transfer are secured, ensuring integrity and confidentiality of data; system monitoring, and incident detection and reporting, methods are implemented; and forensic analysis of cyber incidents is carried out; and

k) Collaborate in the development of ICAO's cybersecurity framework according to a horizontal, cross-cutting and functional approach involving air navigation, communication, surveillance, aircraft operations and airworthiness and other relevant disciplines."

2. Finally, the Assembly Instructs the Secretary General to:

"a) develop an action plan to support States and industry in the adoption of the Cybersecurity Strategy; and

b) continue to ensure that cybersecurity matters are considered and coordinated in a crosscutting manner through the appropriate mechanisms in the spirit of the Strategy."

ICAO Convention on International Civil Aviation Annexes

A number of documents discussed in this chapter refer to ICAO Annexes 17 and 19 to the Convention on International Civil Aviation. Annex 17, titled "Security – Safeguarding International Civil Aviation against Acts of Unlawful Interference," is one of a series of ICAO publications known as "SARPS," or International Standards and Recommended Practices. The last publication of Annex 17 was in March of 2020.

"Since its publication, Annex17 has been amended ten times in response to needs identified by States and is kept under review by the Aviation Security (AVSEC) Panel. The aviation security specifications in Annex 17 and the other Annexes are amplified by detailed guidance material contained in the *Security Manual for Safeguarding Civil Aviation Against Acts of Unlawful Interference* which was first published in 1971. This restricted document provides details of how States can comply with the various Standards and Recommended Practices contained in Annex 17. The Manual has since been developed for the purpose of assisting States to promote safety and security in civil aviation through the development of the legal framework, practices, procedures and material, technical and human resources to prevent and, where necessary, respond to acts of unlawful interference."

As indicated, the Security Manual is a restricted document, meaning that only those individuals or organizations that have ICAO's permission can access the document (another simple example of cybersecurity, in that one seeking access to the document on the ICAO website must have acceptable digital credentials).

Annex 17, in its last publicly available version in 2011, does not mention cybersecurity, but cyber attacks would clearly fall under the definition of "acts of unlawful interference" with aviation assets and infrastructure.

Annex 19, Safety Management, first adopted in 2013, is also a "SARP." This Annex also does not deal directly with cybersecurity. It generally and specifically describes the standards and recommended practices for a civil aviation organization to stand up and manage a Safety Management System (SMS). The document provides top-level guidance to Member States on creating a State safety program, setting up a safety management system, how to organize and manage a safety data collection and preservation system, and outlines the elements of a safety oversight program. Many of the documents and materials discussed in this chapter, as well as some listed in "Further Reading," refer to Annex 19 for a framework in establishing a cybersecurity management system.

The European Union Aviation Safety Agency also embraces the standards and practices set forth in Annex 19.

Other organizations that have addressed cybersecurity and cyber resiliency, publishing documents too numerous to cover here, are listed at the end of this chapter. Some of the better-known leaders in this domain are the International Civil Aviation Organization (ICAO), European Aviation Safety Agency (EASA), The Federal Aviation Administration (FAA), National Aeronautics and Space Administration (NASA), National Institute of Standards and Technology (NIST), International Organization for Standardization (ISO), International Electrotechnical Commission (IEC), Civil Air Navigation Services Organisation (CANSO), and Center for Strategic and International Studies (CSIS).

Conclusion

Cybersecurity and cyber resilience concepts and methods are of critical importance to the continuing success and relative safety of the international civil aviation system. The worst nightmare for governments, aviation authorities, law enforcement, policy makers, users of the system, and the general public is for hostile actors to carry out another terrorist attack like the tragedy of 11 September 2001 in the US. While that event was horrifying for those who were there and anyone who viewed the videos, that scenario could be repeated, several orders of magnitude greater, by a global cyber attack. While there are many layers of protection from such attacks available to governments and organizations, they still happen, as witnessed by more recent "ransomware" attacks that for now appear to be weapons of extortion. User or organizational carelessness is probably the root cause of many of these successful attacks. State sponsored terrorism and cyber events are, unfortunately, facts of life around the globe.

The visionaries and developers of new airspace management concepts such as UTM and U-space rely entirely on the ability to interconnect the ATM system with users, employing digital electronic communication. A wide-area compromise of that system could bring the whole system down, or a focused part of it. One does not have to think too hard to imagine the chaos that would result from having dozens, hundreds, or thousands of manned and remotely piloted or autonomous aircraft occupying the same block of airspace with no way to communicate position, altitude, or speed to anyone else. And with urban air mobility or advanced air mobility, some of those remotely piloted or autonomous aircraft could be occupied by people.

Thus, it can safely be argued that without attack-proof cybersecurity methodology, the ATM system will never be able to maintain or exceed its current target level of safety. Cyber technology is a wonderful thing that has brought much good to the world, but it is also extraordinarily complex, and persons, governments, or organizations with evil intentions are still finding their way into places where they do not belong. After all, the NSA and CIA in the US were themselves the victims of data breaches by unauthorized individuals or entities. If those organizations can be hacked, then no one is safe.

Cybersecurity may be the most urgent topic of this book. Many organizations and governmental entities are working very hard to make internet users safe from attack. Some of the frameworks and concepts of operations to achieve that goal have been outlined in this chapter, but the inquiry should not stop here. More resources and global collaboration would be welcome.

References

Langley, M.E., Sandia National Laboratories, and Sandia LabNews Sandia brings best cybersecurity minds to bear in National Policy Talks, August 2, 2021. Available at: https://www.sandia.gov/labnews/2021/08/02/sandia-brings-best-cybersecurity-minds-to-bearin-national-policy-talks.

A40-10: Addressing Cybersecurity in Civil Aviation. Available at: https://www.icao.int/cybersecurity/Documents/A40-10.pdf#search=A40-10

Administrative package for ratification of or accession to the convention on the suppression of unlawful acts relating to international civil aviation (Beijing Convention, 2010). Available at: https://www.icao.int/secretariat/legal/AdministrativePackages/Beijing_Convention_EN.pdfsearch=beijingconvention

Axios Events. Data security in a hybrid world, broadcast December 9, 2021. Available at: https://www.axios.com/axios-event-data-security-in-hybrid-world-ddd012a1-fac4-4f75-beaf4a9c8a24da9e.html

CANSO. Industry perspective on aviation cybersecurity. Available at: https://www.icao.int/ESAF/Documents/meetings/2021/AFIMidCyberSecurityWebinar9thDecember2021/Update/Presentations/ICAOCybersecurityWebinar-CANSO.pdf#search=cansoairtrafficmanagementcybersecurity

CANSO. Air traffic management cybersecurity policy template checklist. Available at: https://www.icao.int/NACC/Documents/Meetings/2021/CANSO02/D03-CheckList.pdf.

CANSO. Cybersecurity and risk assessment guide. Available at: https://www.icao.int/cybersecurity/SiteAssets/CANSO/CANSOCyberSecurityandRiskAssessmentGuide.pdf#search=cansocybersecurityandriskassessmentguide

Center for Strategic and International Studies (CSIS). Available at: https://www.csis.org/topics/cybersecurity-and-technology/cybersecurity and https://csiswebsite-prod.s3.amazonaws.com/s3fspublic/211105_SignificantCyberIncidents.pdf?_Bux.NVhaioSPTAcspLrKuLx.xCZNSP3

EASA. Opinion No. 04/2020. Available at: https://www.easa.europa.eu/downloads/121096/en.

EASA. ICAO Annex 19 Safety management. Available at: https://www.easa.europa.eu/sites/default/files/dfu/ICAO-annex-19.pdf.

ICAO. Global resilient aviation network concept of operations. Available at: https://www4.icao.int/ganpportal/trustframework.

ICAO. Cyber security and resilience working paper A40-WP/427. Available at: https://www.icao.int/Meetings/a40/Documents/WP/wp_427_en.pdf#search=A40-WP/427.

ICAO. Declaration on cybersecurity in civil aviation. Available at: https://www.icao.int/cybersecurity/Documents/DECLARATIONONCYBERSECURITYINCIVILAVIATION.pdf#search=declarationoncybersecurityincivilaviation

ICAO. Update on the progress in cybersecurity. Ghana, 2019 APIRG/22-IP/11. Available at: https://www.icao.int/WACAF/Documents/APIRG/APIRG22/IPsFINALENG/IP10Updateontheprogressincybersecuritypdf#search=APIRG/22-IP/11

ICAO. 224th Session, Fourth Meeting Mechanism to address cybersecurity in ICAO. Available at: https://www.icao.int/about-icao/Council/CouncilDocumentation/224/CDEC/C.224.DEC.04.EN.PDF#search=C-DEC224/4

ICAO. Thirteenth Air Navigation Conference, Cyber Resilience AN-Conf/13-WP/27. Available at: https://www.icao.int/Meetings/anconf13/Documents/WP/wp_027_en.pdf.

ICAO. CANSO Cyber security and risk assessment guide. Available at: https://www.icao.int/cybersecurity/SiteAssets/CANSO/CANSOCyberSecurityandRiskAssessmentGuide.pdf#search=canso.

ICAO. Annexes booklet. Available at: https://www.icao.int/safety/airnavigation/NationalityMarks/annexes_booklet_en.pdf.

ICAO. Air Traffic Management Security Manual. Available at: http://www.aviationchief.com/uploads/9/2/0/9/92098238/icao_doc_9985_-_atm_security_manual_-_restricted_and_unedited_-_not_published_1.pdf.

ICAO. Air traffic management cybersecurity policy template checklist. Available at: https://www.icao.int/NACC/Documents/Meetings/2021/CANSO02/D03-CheckList.pdf#search=AirTrafficManagementCybersecurityPolicyTemplateChecklist.

ICAO. Doc 9750 Global air navigation plan. Available at: https://www.icao.int/airnavigation/IMP/Documents/Doc9750-GlobalAirNavigationPlan.pdf#search=doc9750.

ICAO. Secretariate study group on cyber security. Available at: https://www.icao.int/cybersecurity/Pages/Working-Groups.aspx.

ICAO. Trust framework update. Available at: https://www.icao.int/APAC/Meetings/2020CNSSG24/SP01_ICAOAI_11.2-ICAOTrustFrameworkUpdate.pdf#search=ICAOTrustFrameworkUpdate.

NIST. Framework for improving critical infrastructure cybersecurity. Available at: https://www.nist.gov/publications/framework-improving-critical-infrastructure-cybersecurity.

Observations from the front lines of threat hunting: A 2018 Mid-year Review from Falcon Overwatch. Available at: https://go.crowdstrike.com/rs/281-OBQ266/images/Report2018OverwatchReport.pdf

The Thales Group. Cyber-security in air traffic management presentation to ICAO, Mexico, 2018. Available at: https://www.icao.int/NACC/Documents/Meetings/2018/CSEC/P05CybersecurityPresentation-THALES.pdf

World Economic Forum. The Global Risk Report 2021. Available at: https://www.weforum.org/reports/the-global-risks-report-2021.

Index
